"You Look Exactly Like A Fresh Rose, Do You Know That?"

Sarah hadn't the strength to reply; all she could do was stand transfixed, her gaze locked to his.

His mouth came down on hers with the suddenness of a hawk diving for prey. There was soon nothing in the world but his hard, lean body against hers, soft and yielding.

"What gave you the right to . . ." she sputtered.

"The right of a man with a maid, my lovely red rose. I'm sure a smart girl like you can work it out."

"Beast!" Sarah shouted.

TESS OLIVER
has received world-wide recognition for her endearing and lively romantic fiction. Her interest in horticulture and music brings to her novels a sensuous richness that is especially appealing.

Dear Reader:

Silhouette Romances is an exciting new publishing venture. We will be presenting the very finest writers of contemporary romantic fiction as well as outstanding new talent in this field. It is our hope that our stories, our heroes and our heroines will give you, the reader, all you want from romantic fiction.

Also, *you* play an important part in our future plans for Silhouette Romances. We welcome any suggestions or comments on our books and I invite you to write to us at the address below.

So, enjoy this book and all the wonderful romances from Silhouette. They're for *you*!

P. J. Fennell
President and Publisher
Silhouette Books
P.O. Box 769
New York, N.Y. 10019

TESS OLIVER

Red, Red Rose

Published by Silhouette Books New York

SILHOUETTE BOOKS, a Simon & Schuster Division of
GULF & WESTERN CORPORATION
1230 Avenue of the Americas, New York, N.Y. 10020

Distributed by Pocket Books

ISBN: 0-671-57014-5

First Silhouette printing June, 1980

10 9 8 7 6 5 4 3 2 1

Printed in the U.S.A.

For Karl
who opened all the doors
to the rooms of my life

Chapter One

In the obscuring grey mist of a Northern California winter day, only a searching eye would have seen the weathered redwood sign leaning companionably into the closely surrounding thicket of bush, hedge, and sapling trees growing down to the edge of the gravel road. Sarah Halston geared down and headed her buttercup-yellow Datsun Hatchback up onto the slight incline of the rutted drive leading to her destination. After negotiating a horrendous pothole, she glanced at her wristwatch and saw that she was fifteen minutes early for her appointment with one Kight Ramsey.

Wanting to put her best foot forward, because a job of this kind was so important to her, Sarah had allowed more than enough time to drive the twenty miles from Sacramento to this small five-acre country estate on the banks of the American River, in the foothills of the Sierra Nevada. Only a few more miles' drive would have brought her into the heart of California's Mother Lode country, where the great gold rush of 1849 began.

Sarah reached the top of the incline, where the drive took an easy, graceful curve which brought her up onto a plateau where the sudden impact of the house caused a smile of delight to soften her serious mouth. She guessed that it had been built about twenty-five years ago, but by an architect who'd been far ahead of his time. The house gave the impression of being a mountain retreat in that it was a comfortably rambling multi-leveled redwood affair with heavy cedar shakes on the roof. But a more careful look revealed that the several roof levels and the proportion of glass to wood were designed by a very sophisticated mind. The grounds surrounding the house, however, were another matter entirely.

The drive continued in a circular sweep leading toward the discreet front door, then back onto itself and down the road Sarah had just traveled. She pulled over to the gravel parking area to the right of the house and stopped her car in front of the open door of a three-stall garage. An enormous white Lincoln Continental and a sleek black Porsche were parked in two of the stalls. The third was jammed full of disreputable garden machinery in various stages of rusty disrepair. Sarah got out of the car to stretch her long legs until her watch told her she was expected.

Even though it was a bleak January day, with the temperature in the mid-forties and a threat of rain hanging in the air, no California landscape should have looked as woebegone and neglected as these grounds did. The hedges skirting the lawn area in front of the house were choked and tangled together, and even from this distance Sarah could see patches of dead branches among the healthy green that should have been pruned out. She rested her cool, critical grey eyes on the lawn area which had been let run to meadow.

"All very well and good," she muttered to herself, "even appropriate for a country house, but weeds are not the same as wildflowers and grasses."

Across the lawn, to the left of the house, was a large rose bed. Sarah glanced at her watch, which told her she had seven minutes more to wait, then glanced at the house to see if perhaps her arrival might have been noticed and her presence required, early though she was. But the house showed no sign of life, so Sarah gave in to her curiosity and walked quickly across the spongy, ill-kempt lawn to the rose bed.

Never in her three years as a jobbing gardener had she seen a formal rose bed allowed to reach such a scandalous state. The bushes looked as if they hadn't been pruned for several years, their centers an angry-looking mass of twiggy growth and awkward doglegs. The tattered leaves hung on like dirty rags and brown wizened buds drooped down from weak cane tips.

Inasmuch as rose bushes bloomed through Christmas and surged into spring growth as early as February in this mild climate, most rosarians arbitrarily forced their bushes into at least a short dormancy by pruning no later than the last of December. If these bushes weren't pruned within days, it would be too late to prune at all this year.

Sarah spied a thick cane growing at an angle Nature herself would never have attempted. Looking more closely, Sarah saw that the cane had broken jaggedly between two nodes. If a clean cut weren't made soon, the cane might be lost entirely to die-back. Of course, it was none of her affair. Yet. Or ever, perhaps. On the other hand, how could anyone object to her just popping into the bed and making a quick cut? Surely no one who had a rose bed in the first place could consider such an act of mercy as trespassing, could they?

One quick involuntary glance at the house showed her that a part of her mind was not entirely convinced she was doing the right thing, but the broken cane bothered her in the same way an unset bone might bother a nurse. She pulled out the clippers from the canvas pouch that served as dual small tool receptacle-and-workaday handbag. As she wormed her way into the bed, the rose thorns caught on her heavy denim jeans and jacket. When she reached the bush she saw that the cane was, in fact, a sucker, growing from beneath the bud union. "Talk about trespassing," she said sternly. "Sorry, but out you come." She knelt down and, while holding herself up on one elbow, lowered her torso in a concave arch so that she could get at the base of the bush without being impaled on its thorns. She worked away at the thick sucker, her bottom thrust into the air, and as so often happened when she worked, she forgot time and place as the absorbing nature of her work led her from one cane to another, one bush to another. She might have gone on for hours if a deep voice hadn't suddenly boomed out behind her.

"This property is already famous for its view. If you keep this up, we'll have to put up an electric fence to keep out the public!"

Startled, Sarah whirled around to see who the voice belonged to, catching her thick auburn hair on a thorn. In embarrassed panic she tried to wrench her hair free and caused the thorn to bite deep into her scalp and scrape down the side of her temple. With every jerky move she made to extricate herself, her clothes were caught anew by thorns and stiff canes on every side.

"Stand still, woman! Stop thrashing around and let me unloosen you," the man responsible for her mortification said. Stepping gingerly into the rose bed, he untangled her clothes. Then, holding her by the hand, he guided her out to stand beside him on the lawn. "Are you all right?" he asked. And then: "Good Lord, you're bleeding."

Sarah's right hand fluttered absurdly as it tried to smooth her hair and her clothes, as if it felt off balance without its twin, which the man still firmly held. Retrieving it from him and then suddenly not knowing what to do with either of them, Sarah said with a nonchalance she was far from feeling, "I'm perfectly all right. It's nothing."

He seemed awfully tall to Sarah, who was a tall girl herself and used to being nearly on eye level with most men. With him, however, in order to look up into his dark, shadowy eyes, she had to tilt her head quite far back. Or was that only because he stood so unnecessarily close to her? Whatever the reason, her glance was very brief. It seemed that the combination of the scare he'd given her and the unexpected stinging pain were making her feel giddy.

The man before her, scrutinizing her with concern, and yet with repressed amusement, too, was dressed in a black-and-red-checked wool shirt and a pair of faded jeans. His clothes and the sturdy rawhide boots he wore had all seen yeoman service. He took a clean handkerchief from his back pocket, and with startlingly strong fingers he took hold of her jaw to hold her face steady while he moved to wipe the blood trickling from her temple down the side of her face. Blushing from what suddenly struck her as an uncalled-for intimacy, Sarah freed her face from his hand and frowned slightly. "Really, it's nothing," she said coolly. "Please don't fuss."

He backed up a step and held his hands palm up in mock surrender. In a mild, amused voice, he said, "I do beg your pardon. It's just that I feel responsible for your wound."

"And so you should," she retorted contrarily. "Prowling around like that, making fresh remarks, scaring people half to death."

"I'm sorry I startled you, but as for prowling around, isn't that the pot calling the kettle black?"

Sarah had quite forgotten her own part in bringing about this fiasco, and what with one thing and another, she didn't thank this man for reminding her of it. Where in normal circumstances Sarah, a fair-minded girl, would have willingly admitted her share of the blame, now some disgruntled imp inside her suggested that the best defense was an offense.

"I never would've done that if this rose bed weren't in such shocking condition. If taking care of this garden is part of your job here, I can certainly understand why Mr. Ramsey is advertising for a new gardener."

His heavily fringed eyes opened wide in surprise, a large smile spread over his well-defined mouth. "You're a gardener?" And then he tossed back his well-sculptured head in rich, rolling laughter. Sarah's heart cringed in dismay at his ridicule. Of all the various reactions she habitually received to her choice of work, it was, strangely, amusement that hurt her the most. It seemed to make a fool of her, to denigrate her more than the harsher reactions of outrage or contempt ever did. At least these people took her seriously. Mere disapproval was the *best* she could ever hope for. She knew it was silly to mind being made light of, and she was working hard to learn to ignore it, but for a girl who valued her dignity as much as Sarah, it wasn't easy.

As his laughter pealed out into the grey misty air, she couldn't avoid staring mesmerized at the strong column of his exposed neck, her eyes wandering to the fine breadth of his chest and shoulders, the muscles obvious even under the thick but close-fitting shirt. His laughter wound down finally, leaving only a remnant of a sweetly sly smile on his disturbingly handsome face. He shook his head as if to clear it and a dark lock of hair slid enchantingly over one eye. As he scooped it back she saw, too late, that he could never be a gardener with those strong, masculine, but very well-cared-for hands. She thought uncomfortably of her own grubby, embattled-looking hands, the nails unpolished and cut short for efficiency. She really must try again to work with gloves, she thought distractedly.

"No, I don't do the gardening around here," he said, at last responding to her question. "A good gardener is an artist, and I wouldn't presume," he said with a charming smile that partly mollified her injured feelings.

"All is forgiven," she said with an answering smile.

"For which I'm very grateful," he said in an oddly serious tone.

"Well, then, since you 'don't presume' to be a gardener," Sarah said in a mock-pompous tone, "what is your job here?"

He shrugged. “A little of this and a little of that.”

“General overseer? That sort of thing?” she pressed, for she wondered how much of this provocative man she might see if she were to be hired by Mr. Ramsey.

“Yes, that about sums it up,” he answered as he walked her toward the house. When she realized that he was heading for the front door she hung back, an all too clear picture of her dishevelment taking form in her mind. She took a quick look at her wristwatch and saw that it was only a few minutes after the hour. What a lot had happened in seven minutes, she marveled.

“I wonder . . . I don’t want to keep Mr. Ramsey waiting,” Sarah said hesitantly, “but perhaps I should freshen up a bit?” She looked down at her soiled hands and damp knees and thought ruefully of what a fine impression she’d make now on Mr. Ramsey, totally discombobulated and dripping blood as well. It had suddenly become even more important than before to make a good enough impression to be hired. Perhaps she oughtn’t to have worn her work clothes for this interview, as was her custom.

“Is Mr. Ramsey a very formal man?” she asked in a worried tone.

The tall man at her side smiled down at her. “Oh, not particularly. Why do you ask?”

“He won’t expect pumps, a wool skirt, and matching sweater set, I suppose?”

“Not on a gardener, I shouldn’t think,” he said gravely.

Sarah sighed in relief. “That’s all right, then. You see, I’m a working person and I take my work as seriously as the next . . . well, man. I’ve found that if I wear my working clothes for the first meeting with a new employer, we get off on the right foot from the start. They seem to understand then that I mean to be taken as just another gardener, you see, sex notwithstanding."

He turned his head away from Sarah slightly and put his loosely fisted hand up to his mouth, ostensibly to cover a not very convincing cough. Once again, Sarah noted his amusement with a twinge of hurt.

"You don't expect me to believe, I hope, that mere work clothes are enough to keep men at a distance from you? If so, you must know only fashion designers."

While the woman in her delighted in the compliment, the working Sarah replied in a dignified tone, "My male employers come to understand that I come to their gardens to work, not to walk down the primrose path with them. And if they don't understand that, it's easy enough to remedy. I simply leave and find work elsewhere."

He nodded once, as when a bargain is struck. "Point taken." They'd reached the front door and Sarah waited for him to ring the doorbell. Instead, he took her by the arm and turned her to face himself. At the touch of his hand, she had a sudden wild wish that she hadn't so idiotically moved away from him a scant few minutes ago. In case she never saw him again after today, in case he never touched her again, it would've been good at least to have that memory of him cupping her face in his hands.

He leaned forward to look closely at the scratch on her temple, and her stomach quivered in reaction to his warm breath on her hair and cheek. "Of course, you'll want to clean up that wound you got in the war of the roses," he joked. "I'll have Mrs. Mole see to it. Don't worry about keeping Mr. Ramsey waiting. Mr. Ramsey was no more prepared for you than you are for him."

"Oh? That must be what you'd come out to tell me in the first place, was it?"

"Come in now," he said, smoothly ignoring her question, and to Sarah's surprise, he opened the door without ringing and entered the house. Well, that was typical California informality. She barely had time to take in the vastness of the entry hall and the fact that it was an atrium full of semi-tropical plants basking in the opalescent light that poured from the ceiling, before she was introduced to a sour-looking woman who suddenly appeared from a hall off the entry.

An amazing transformation came over the seemingly charming and genial man at the older woman's appearance. His manner was one of cool reserve and his tone curt, even somewhat arrogant, as he said, "This is Miss Halston, Mrs. Mole. She's come to interview for the position of gardener. She's had a slight accident, as you see. Take her to the bathroom and see that she has everything she needs. When she's ready, show her into the study."

Mrs. Mole pursed her lips in an expression of disapproval that Sarah had seen many times before. Nodding curtly, she turned on her heel so quickly to leave the room that Sarah had time for only the briefest glance over her shoulder at the man before scrambling after her, lest she be left behind. She followed the stiff back of the older woman down the hall to a small, elegant bathroom. Mrs. Mole silently handed her a box of plastic bandages and a face cloth, whereupon she closed the door behind herself, leaving Sarah to only hope that she'd make herself available, as she'd been asked, to show Sarah into Mr. Ramsey's presence.

Sarah discovered that there was no way to make the plastic bandage stick to her wound since it was on her scalp, and no need, anyway. As she dabbed at the drying trickle of blood on the side of her face, she realized how really gruesome she must have looked to . . . but what was his name? It suddenly struck her that if she wasn't hired here, and she didn't see him before she left, she might never know his name. Certainly she couldn't ask Mrs. Mole. Not for anything would Sarah make herself even slightly vulnerable to that lady! At the thought that she might never see him again, her heart squeezed painfully and she was appalled at the strength of her own feelings. It wasn't at all like Sarah to be carried away by emotional whims. She'd met the man a scant fifteen minutes ago, and here she was carrying on as if they'd been close for a lifetime.

Ridiculous! she thought, scrabbling in her canvas pouch for her hair brush. She pulled it roughly through her medium-length auburn hair a few times and tucked under the ends here and there until it fell into the burnished pageboy style that she thought best suited her face. Not bothering with lipstick—after all, she was a gardener, not a fashion plate—she took a deep breath and opened the bathroom door to find Mrs. Mole standing sentinel just outside. With a curt tilt of her head, the woman indicated that Sarah was to follow her.

Sarah was shown into a room with a glass wall looking out over a panorama that explained the handsome man's remark about the fame of the estate's view. Directly outside the sliding-glass door was a large brick terrace furnished with redwood patio furniture. Beyond it she saw a pool, a large poolhouse, and beyond that a tennis court. But what drew the eye was a wide unhampered view of the beautiful grey-green American River flowing serenely between its wooded banks. Sarah could imagine that the sight would be made even more spectacular at night by the twinkling lights of the city of Sacramento beyond the far bank.

The walls of the room itself were paneled with warm redwood bookcases and display shelves. Its furnishings were contemporary: a straight-lined beige suede sofa, chrome and glass tables, a sleek desk of glowing teak covered with what appeared to be blueprints. A luxuriously deep-piled carpet in a subtle buff, brown, and terra cotta native Indian design covered the center of the dark brown hardwood floor. There were models of buildings here and there on the display shelves, some of them vaguely familiar to Sarah, and she was just about to take a closer look at them when she sensed that someone had joined her in the room.

She turned to see the man she'd worried she'd never see again and her heart lifted in gladness. He'd changed his clothes and now wore a pair of camel's-hair slacks, a cocoa-colored V-neck cashmere sweater with an open-necked grey silk sport shirt worn underneath. His rough boots had been exchanged for an expensive-looking pair of brown leather loafers.

With a warm smile, Sarah started toward him. "I was just thinking, we haven't even introduced ourselves. I'm Sarah Halston, and you're . . . ?" And then it all began to come clear to her. The smile faded from her face like a wilted flower. "But you knew who I was. You introduced me to Mrs. Mole. You came in the house without ringing. You must be . . ."

"Yes, I'm Kight Ramsey." He smiled impishly, as a small boy might smile if caught in some mildly naughty behavior by a mother he knew doted on him.

Sarah went cold all over. Shame for the yearnings she'd felt toward him warred with a building anger at the callousness with which he'd played her for a fool, humiliating her, laughing secretly at her because she'd so stupidly mistaken his identity, so girlishly confided to him her uncertainties about the appropriateness of her clothing. From behind a thick curtain of outrage she was dimly aware that he seemed to consider the matter closed and was talking about the position, the duties it entailed, and so forth.

"Very large area to take care of,"—"perhaps too much work for one person"—"blank check and all the extra help you might require . . ." She heard these remarks in snatches and patches only as she stood frozen, paralyzed by a welter of emotions more intense and painful than she'd felt in years.

She suddenly heard her own voice croak out, "How could you do such a thing?"

The insouciant smile left his face to be replaced by a look of surprise, and that in turn by one of cool caution. After a moment's dead silence, he broke eye contact with her and said with aloof politeness, "I beg your pardon. Of course I was wrong to deceive you. It's just that I really didn't know who you were at first, and by the time I realized, it was too late."

"How so, too late?" she demanded. "Too late to get out of it without embarrassing your precious self? Too late to backtrack to the role of dignified employer once you'd behaved like a brash street urchin?"

"Now, just a minute, Miss Halston, you're not entirely without—"

"I have nothing further to say to you, Mr. Ramsey," Sarah interrupted bluntly, moving toward the door of the room.

"Quite the kangaroo court you've set up here," he said, reaching out as she passed him to grip her forearm in steel fingers. "Behold Sarah Halston: accuser, judge, and jury, all wrapped up in one very pretty package. Oh, there I go again, committing the crime of remarking on your appearance," he said with hateful sarcasm.

Sarah tried to yank her arm away but, embarrassingly, couldn't move it even an inch. Cursing herself for ever wishing to feel his touch on her face, she said in what she hoped was an insultingly patronizing tone, "Mr. Ramsey, don't flatter yourself that you're a criminal. It's only that your observations are pathetically irrelevant. If I'd come here to interview as a model, my looks would be pertinent. As a gardener, they simply aren't. Now, as I said, I have nothing further to say to you, so please take your hands off me."

In response he only tightened his grip until Sarah began to actually be frightened of him. After all, what did she know about this maniac? When he spoke his voice was like ice.

"Because you have nothing to say to me, it doesn't follow as the night the day that I have nothing further to say to you, Miss Halston. And furthermore, I'm not accustomed to having a prospective employee decide when a meeting with me is adjourned."

"How wonderfully feudal for you, Mr. Ramsey, but

your lordly expectations don't signify here because I am not a prospective employee of yours. I wouldn't work for you if I were on the verge of starvation."

"You're quite a speech maker for someone who has nothing to say," he drawled sarcastically. "But as long as I'm physically stronger than you, I intend to get just a few remarks in sideways between your hysterical diatribes. First of all, you're not entirely without blame for this debacle. Didn't you hide your actual identity from me by concealing that you were a woman? How could I know that when your letter of reply to my advertisement was signed simply 'S. Halston'?"

"Even you," she said coldly, "must see that most employers wouldn't even grant me an interview for such heavy work as gardening if they knew I was a woman. It would be an impossible handicap that I'd be a fool to labor under."

In the seconds while he considered her reply, Sarah tried again to free her arm from his painful grasp and was further humiliated when he didn't even seem to notice her puny effort.

"That's quite a large generalization, isn't it? Doesn't it amount to the same sort of prejudice you accuse others of?" he said smugly.

"Let me just ask you this," Sarah replied. "If you'd known I was a woman from my letter, would you have granted me an interview?"

A flicker of uncertainty loosened the angry set of his face. His voice was thoughtful as he answered, "I don't really know. And now we'll never find out, will we?"

His words touched something soft and tender deep inside Sarah. There was so much now that they'd never know, she thought sadly. Slowly, as if regretting to break contact, he loosened her arm from his grip. Or perhaps the regret was in Sarah, not him. Her feelings were in too much turmoil to be sure. With what now felt more like sorrow than anger, Sarah turned toward the door when he spoke again, in a voice more compelling than even his touch had been. "May I say one last thing, Miss Halston?"

Sarah hesitated, keeping her back toward him, lest she throw the last shreds of her self-respect to the wind at the sight of him, then nodded reluctantly.

"I ask you to remember that I did admit my manners were poor and that I asked your pardon," Kight Ramsey said quietly. "Now I ask you to forget this bad start we've had. Perhaps you don't know that I've only recently bought this place. It was the previous owners who let the grounds go to wrack and ruin. As you see, they desperately need the attention of a dedicated gardener, and I feel certain that's what you are—sex notwithstanding." He dryly stressed the last words to remind Sarah of her own words. "And, after all, that was the original intention of us both, wasn't it? Mine to hire a gardener, yours to be hired? Please say you'll take the job."

Sarah realized that he thought the bone of contention had to do with his treating her as a woman as distinct from the way he'd have treated a man who'd come to be interviewed for the position of gardener. And she acknowledged that both of them had somehow got blown onto that track—the wrong one though it was. No, Sarah wasn't a fanatic, either in her life or her work. She was glad to be a woman, and she'd learned to tolerate the usual male response to her femininity that she usually received at first. But what would be the good of belaboring all that now? And yet . . . and yet, perhaps she owed at least a token explanation, if not to him, at least to her own self-respect.

She drew a deep breath and turned to face him. The hopeful expectancy of his mien called forth from within her such a strong desire to satisfy it, to say yes to him, that she very nearly lost all control. But no, she mustn't let herself be taken in again. No good could ever come from such a poisonous beginning.

"Thank you for your apology, Mr. Ramsey, and I, too, am sorry this hasn't worked out. But you've made me feel like a prize fool, and I think you wouldn't want to employ a fool."

Kight Ramsey flinched from her words and a dull, unhealthy color suffused his strong face. Turning away from Sarah in what anyone would have recognized as a dismissal, he muttered in a tight voice that seemed meant more for his own ears than hers, "It's clear enough that there's a fool in the room, all right."

Sarah turned for the final time toward the door and saw Mrs. Mole standing just inside the room. Heaven only knew how long she'd been there. She held in her hands a tray of coffee and what might have been pound cake or a fruit bread; Sarah couldn't see it clearly for the tears in her eyes. What she could see was that Mrs. Mole's earlier scowling expression had changed to a creamy, gratified smile.

Chapter Two

In the first few groggy minutes after Sarah awoke the next morning, she'd gladly have signed away everything she owned, as little as it was, just for the right to stay in bed all day. She felt depressed and dull in both mind and spirit. But Sunday was the day she spent with Iris Millidge, and for nothing short of major surgery would Sarah disappoint darling Iris. She threw aside the bedcovers and padded on bare feet into the kitchen, where, as was his wont, Duke sat imperiously on the middle of the kitchen table waiting for his breakfast with an expression of strained patience. "Beg pardon for sleeping late, most illustrious eminence," Sarah murmured as she opened a fresh can of food and spooned out his morning portion.

While the large marmalade cat made short work of his breakfast, Sarah put coffee on to brew and an egg to boil, then walked into the main room of her apartment and drew the oatmeal linen drapes that covered the glass wall giving out to her back garden. The apartment complex where Sarah lived had twenty-six units built around a common area in the middle with a patio and small swimming pool. In the back of each unit was a small garden for each tenant to do as much or as little with as he wished. Because the area was so small, Sarah had landspaced hers in the Japanese manner—sparsely but elegantly.

For the hundredth time this year Sarah wished that Ben were here to see the garden develop that he'd helped her design three years ago. Ben Yashimoto, her Aunt Elaine's gardener, had been the main source of affection and guidance in Sarah's young life. At the age of six Sarah had been orphaned when her parents died in a hotel fire while on holiday in France, and thereafter she'd lived with her mother's sister. Aunt Elaine had been kind in a dutiful and distant way, but she was a high-strung, energetic woman not much interested in family life. She preferred to spend her time on the boards and committees of Sacramento's many artistic and cultural organizations, so the actual raising and day-to-day care of the growing child had been turned over to a series of housekeepers who came and went with Aunt Elaine's whims and tempers. It was Ben Yashimoto who'd been the one constant in Sarah's life, with whom she'd spent most of her time when she wasn't in school, to whom she'd confided all those childish problems that most children take to their mother. It wasn't that Ben wasn't fired as regularly as the housekeepers, it was just that when Aunt Elaine dismissed him, in a flare of pettish temper, he bowed ceremoniously to her and simply went back to work as if she'd never spoken. Within the day, usually, she, too, would pretend it had never happened. Aunt Elaine's garden had been important to her as a showpiece, and she knew Ben was virtually irreplaceable.

As a child, Sarah had mistaken her aunt's interest in

the garden for real regard, and it was her desire to please Aunt Elaine by sharing that regard, combined with Ben's influence, that developed Sarah's love and talent for the art and skill of gardening. By the time she was seventeen and nearing high school graduation, she'd worked out what she felt was a sensible plan for her life. She wanted to work at the design and maintenance of gardens: residential, commercial, and institutional. But that meant a degree in Landscape Architecture. Even today, at the age of twenty-four, the memory of Aunt Elaine's incredulous laughter on hearing Sarah's plan brought back the painful flush of humiliation to her face.

"You've always been an odd child, Sarah, but this is the limit! Why, it's insane for any girl to think of doing that kind of work, but for a girl of your background it's absolutely out of the question. How ridiculous you are! Do you think I gave you years of careful upbringing so that you could spend your life doing the work of a servant? So you could become a common laborer? Hardly! I've planned for you to attend a lovely Liberal Arts college in the east. When you've graduated, perhaps you'll find suitable work in a publishing company or a museum until you marry."

Sarah's deepest and most secret desire was to marry and have children that she could love and nurture in all the ways she herself had never known. But because she'd never inspired more than detached affection from her aunt, she'd subconsciously assumed that she was an unlovable person who would never inspire the love of a man, and, like her aunt, would never marry. Thus, she was all the more dismayed and confused by her aunt's reaction to her plans. Feeling the bonds of obligation too strongly to defy Aunt Elaine, Sarah obediently did what was expected of her—up to a point.

But she took every botany course the college offered and every art course that dealt even in passing with landscape design. She corresponded regularly with Ben and he gave her what amounted to a correspondence course in the work she loved and was still determined to do someday. When Sarah was a senior, Aunt Elaine impulsively sold the Sacramento house and went to Paris to live, leaving Sarah without a home. Ben was retained by the new owners and it was to him that Sarah returned after graduation, refusing her aunt's lukewarm invitation to join her in her chic apartment across from the Bois.

Ben helped her find the apartment where she presently lived and worked out an arrangement with the manager for her to pay half-rent in return for light maintenance of the grounds. For two satisfying years Sarah did job gardening under Ben's guidance, soon earning enough money to put some aside for her still strong ambition to go back to school for her degree in Landscape Architecture, or perhaps to start her own landscaping and maintenance company.

And then, a little over a year ago, at Christmas time, disaster struck. Ben died of a heart attack. Sarah grieved for him and for herself, as well, because in all the world now, there was no one close to her, no one who knew her history, who knew her ambitions and approved them. She found that without Ben's emotional support and encouragement, her ambition was weakening. More and more often during this past year, she'd found herself wishing to find a position as head gardener on an estate, to make that her family, her world, her anchor in life, as Ben had made her aunt's estate his.

As Sarah sat now without appetite over her breakfast, she remembered the wording of the advertisement she'd answered that had led to yesterday's interview: *Experienced head gardener to develop, oversee, and maintain 5-acre country estate near foothills. Must live on premises. Private accommodations provided. Salary negotiable. References required. Reply Box 900.*

To Sarah it had seemed the answer to a prayer. With the permission of several of her clients, she'd mailed their names as references along with her résumé to the box number the following day. On Friday she'd received a call from Mr. Ramsey's secretary, who fixed the time for her appointment and gave her directions to the acreage. The rest was history, Sarah thought ruefully.

If she'd been a superstitious young woman, Sarah might have said the position was bound to come to nothing simply because it was too good to be true. And Kight Ramsey—well, he was too good to be true, as well. If only she hadn't been so personally attracted to him. Then, when he'd played her for a fool, she could have laughed it off as he'd done. After all, those whom we have little feeling for have little power to hurt us. Or—since she had to admit that she'd made herself vulnerable to him, had softened at his touch and confided her fears to him—if she'd not revealed it to him, she might at least have saved face. But as it was, she'd let him see by her subsequent anger that she was attracted to him. For surely, if he thought about it at all, he'd realize that only a personal involvement would cause a reaction as strong as hers had been. And that was an impossible situation for an employer-employee relationship. She could never regain her dignity or maintain her independence under such a circumstance. And in addition, it would be too painful to see him every day knowing that a man like him would never see her as anything but a slightly ridiculous employee. He already thought her a fool, as his last words to her proved.

"Well, so much for empty dreams," she said ruefully, then rose to clear away her breakfast dishes. It was getting on toward noon, nearly time to pick up Iris for their Sunday afternoon visit. She let Duke out to make his morning rounds and went to the bathroom to wash and dress. She'd promised Iris they would go to Sutter's Fort today, one of the little girl's favorite outings, but it was such a chilly day that Sarah hoped to persuade the child to stay in for a quiet afternoon of Old Maid, TV, and cookie baking. Sarah dressed in a pair of black wool slacks and a black and white geometric print cotton blouse that set off her auburn hair and creamy skin to good advantage. She switched on the message receiver attached to her telephone, flung her tan raincoat over her shoulders, and slipped out the door.

Iris Millidge lived in an older area of Sacramento, not far from Sarah's apartment by car, but worlds apart in every other way. Mike and Maggie Reilly, Iris' foster parents, lived in a small rented bungalow in an area robbed of its rightful share of the sun by the giant concrete wings of Interstate 80 that swooped overhead, hurrying its passengers through this shabby, forgotten part of the city and on to the glorious Sierra Nevada and, for three thousand miles, all points east.

Sarah had barely stopped the car in front of the

bungalow when the door was flung open and Iris, barefoot and coatless, danced out onto the front porch waving madly at Sarah, a gamin grin on her homely little face.

"Sarah, Sarah, I'm almost ready! Maggie says come in!"

Sarah entered the small living room, with its well-used furniture and smells of good cooking in the air, to be welcomed by Mike Reilly's bear hug and Maggie's nervous twittering when she caught sight of the long scratch on Sarah's face. "You never take any care for yourself!" she fussed. "Such a beautiful face and hands—and just look how you abuse them!"

"Leave Sarah alone, woman," Mike scolded. "What do you expect when a person does an honest day's work? Get the child ready. Sarah can't wait around here forever."

Maggie threw up her hands in resignation and turned her attention to prodding Iris into her shoes. "You should've been ready and waiting long ago instead of trying on every last stitch in your closet," she fussed at the child in her high-pitched voice.

"Leave off, woman, before you deafen us all," Mike said with meaningless gruffness. The Reillys were in their mid-sixties, both of substantial size, both with grizzled hair. They'd been married so long that they'd come to look alike, and their temperaments, as well, fit hand in glove. Mike was the ballast for Maggie's airborne nature, she the leavening for his stodginess. Because of their devotion to each other and their experience in raising a creditable flock of their own, they'd been especially chosen for Iris, and a highly gratifying choice it had been. Anyone seeing the child a year ago, when first the Reillys and then Sarah had come into her life, would have found it hard to believe that this sprightly, smiling imp was the same wretched little girl of the past.

A few weeks after Ben's death, Sarah saw an appeal in the newspapers from an organization called Friends of Childhood, where the children, all wards of the court, were matched up with people who had not only time, but love, to spare. Sarah'd gone there on a day very much like this one, chilly and wet, and looked at photographs and read case histories one after another until she was sure her heart would break. In private memory of Ben Yashimoto, Sarah chose Iris Millidge to befriend because, in addition to a horrifying family situation, she was such a homely, wan, unprepossessing child that Sarah was afraid no one else would ever choose her.

While Maggie stuffed Iris into her coat, Mike said to Sarah, *sotto voce*, "Young Bill came to see us the other day. Thought our girl was coming along fine." Bill Blanding was the social worker who made up the third arm of Iris' devoted cheering section. "She hasn't had a nightmare for a whole month now," Mike said with huge satisfaction.

"It's all thanks to you and Maggie, Mike. Few people could've worked the miracle you have," Sarah said.

"Go on, you've done as much," Mike said, smiling at her fondly.

Sarah shook her head. "No. You've provided the meat and potatoes. I've just given her an occasional sweet, a once-a-week treat."

"Well, be that as it may, Sarah, the job's not done yet. Our girl has a long way to go. That's what keeps me up nights, though I can't let on to the old woman. Our girl's only seven. What's to become of her when Maggie and I get too old to look after her, or, worse yet. . . ." He left the sentence unfinished.

"Now, Mike," Sarah said sternly, "you're not to worry about that. You know Bill and I will see she's well taken care of, one way or the other."

"That's what worries me, Sarah—that 'other.' Our girl has had enough 'other' in her seven years to break a strong man. I don't think she'd survive again."

There was no more time for private conversation because Iris was at last shod and hatted and coated enough to satisfy even Maggie, and she was handed over to Sarah like a fat little sausage. Amid gay good-byes and numerous kisses, Iris and Sarah made their way to the little yellow car and were soon on their way.

"Why were you trying on all your clothes, darling? Did you have trouble finding something to fit you?" Sarah asked. She kept watch over these matters because the stipend the Reillys received from the agency for Iris' care couldn't really be stretched far enough to meet the growing child's needs, and they had little extra to spend from Mike's modest pension and Social Security checks.

"Oh, no," Iris replied in her clear, piping voice, "I was just playing dress-up."

Seizing the opportunity, Sarah said, "Since it's such a nasty day, would you like to go home with me instead of going to the fort? We can play dress-up and make cookies—whatever you like. We could save the fort for a nice sunny day, perhaps take a picnic."

Iris turned a trusting face to Sarah. "I don't ever care where I go, Sarah, so long as I'm with you."

Sarah's throat swelled with love. "I feel just the same, sweetheart."

When they reached Sarah's front door they found a damp but stoic Duke sitting neatly on the front stoop with his tail curled meticulously around his front paws. Iris swooped down on him and carried him into the house. "I will be the mother and Duke will be my baby. What will you be, Sarah?"

Not for the first time, Sarah noted that the concept of father was understandably repressed in Iris' mind. Thinking it better all around to avoid playing family roles, Sarah suggested circus roles instead, which met with Iris' enthusiastic approval.

So as it turned out, Sarah was a clown. With bright red lipstick she drew large round freckles on each cheek and the tip of her nose, blacked in ferocious eyebrows with liquid shoe polish, and because she had nothing white with which to draw the traditional half-moon mouth, she had instead a lavender mouth, using up an entire tube of eye shadow in the process. Iris' delight spurred her on to such heights that she ran across the common to borrow floppy black scuba fins from a neighbor to put on her feet. With two bed pillows tied front and back under a loose smock, Iris declared that Sarah looked the part so well any circus would be happy to hire her.

Iris herself was a dazzling trapeze artist in a white body suit of Sarah's (never mind if it sagged a bit), topped off with a pink satin evening shell that on Iris's small body looked for all the world like a gorgeous mini-shift. She asked to apply her own makeup, and never before had Sarah seen glamour laid on with such a lavish hand.

The full extent of Duke's grudging cooperation had been to impersonate a trained bear. During most of the fun he'd crouched in the corner of the room, entangled in a fuzzy brown mohair sweater of Sarah's, his amber eyes half-closed with weary scorn at such unseemly carryings-on.

But when the hilarity died down and activity started up in the kitchen, Duke's basic magnanimity came to the fore. Forgiving the humans their temporary loss of good sense, he wriggled out of his costume on the way to his post in the middle of the kitchen table, where he waited with confidence for his fair share of the chocolate-chip cookie dough which happened to be one of his particular favorites.

Sarah was putting the first batch into the oven while Iris, standing on a chair to bring her to counter level, was spooning a second batch onto another cookie sheet, when the doorbell rang. Before Sarah could react, Iris jumped down and ran from the kitchen. Sarah, with a rolling waddle because of the scuba flippers, followed after her. When she reached the living room, Sarah stopped dead, a full room away from the astounding sight of Kight Ramsey, filling up the doorway and dwarfing the child, who craned her head up to see his face.

"Oh, no," Sarah whispered. And then as the full realization of what she must look like made its impact on her, a low involuntary moan escaped from between her lavender lips. With horrible clarity she saw his warm brown eyes open wide and his mouth twitch once before he tightened it to smother a smile. Iris stood looking from one adult to another, then broke away and ran to throw her arms around Sarah's legs. Looking up with a worried frown on her small face, she whispered, "Who is that man, Sarah?"

Sarah bent down to put a reassuring arm around Iris's shoulders. "It's all right, dear. He's just a man I know. Why don't you run along and finish putting the cookies on the sheet? When the oven timer rings, you come and tell me. Don't you open the oven yourself, do you hear?" With a small nod and one last doubtful look at the very large stranger, Iris ran back into the kitchen.

Hoping to offset her ridiculous appearance, Sarah called up every last ounce of dignity she possessed and said in a coldly polite tone, "I was under the impression that we had nothing further to say to each other. Since you seem to think otherwise, you might at least have called first to see if a visit was convenient. But then, other people's feelings aren't number one on your list of priorities, are they?"

He lowered his eyes, Sarah noted, seeming to find her carpet a fascinating object—no doubt so that he wouldn't kill himself laughing at the sight of her. "I'm sorry, Miss Halston, to owe you yet another apology in such a short span of time. But the fact is, I did call first. I left a message on your answering machine."

"Oh," Sarah said flatly, glancing toward her phone with a frown, as if it had betrayed her.

"Shall I repeat the message now? Or perhaps you'd rather hear it for yourself to make sure I'm telling the truth. I know you have little faith in my veracity."

Sarah made an exasperated sound. "Don't be silly. I haven't called you a liar. It's just that I'd planned to listen to my messages later this evening, after I'd taken Iris home."

"Oh," he said with interest. "Then the little girl isn't yours?"

"Of course not," Sarah snapped.

He raised one dark, silky eyebrow into a questioning arc. "Why 'of course'? You look like you'd make a wonderful mother. I can't remember when I've seen such a winsome domestic picture as you've shown me this afternoon—play acting, cookies, a nice, cozy home on a cold, rainy afternoon. . . ."

Offended at what she took to be his amused ridicule, and the arrogance he showed by poking fun at a situation about which he knew nothing, Sarah drew herself up and replied in a lofty voice, "I have *shown* you nothing, Mr. Ramsey; you've intruded where you weren't invited. Please say what you've come to say and then take your leave."

From the look that came into his eyes and the sobering of his expression, Sarah saw—not without a twinge of regret—that her words had found their mark, had paid him back in some measure for his too flippant disregard for the feelings of others. When he spoke, his voice also had lost its wry lightness and become aloof—the same voice he'd used to speak to Mrs. Mole.

"I came to ask if you'd reconsider and decide to come and be head gardener of my grounds. But now I see that you will not. Instead, I'll change my request and ask if you'd be willing just to put the rose bed back into good condition. You gave me the impression that it should be seen to right away, and, after all, you did make a start on it, didn't you?"

Sarah looked closely to see if he was making fun of

her again, but there wasn't a trace of humor on his lean-jawed face, nor a trace of warmth in his dark eyes.

"No, Mr. Ramsey, I don't care to work for you on either a full-time or a part-time basis," she said stiffly.

He shrugged his wide shoulders expressively and nodded, as if her answer were only what he expected. He turned toward the door, pulling his raincoat up closer around his neck, where, Sarah noticed, it disarranged soft tendrils of dark hair.

"A typical female response," he said with a sardonic smile, "but somewhat surprising from the mouth of a gardener who takes her work so seriously."

"And just what do you mean by that?" Sarah demanded.

"Women are so apt to get hysterical over petty matters, aren't they? To think with their glands instead of their minds? How many men would've got in such a snit, do you think, over so small a matter as our . . . little misunderstanding yesterday?"

Outraged, Sarah cursed herself for feeling she'd been too harsh. Imagine the gall, the effrontery! Just as she was about to open her mouth to retaliate, she suddenly got the distinct impression that he was waiting—lying in wait!—for her to do just that. *Hold on,* she said to herself, *you've made a fool of yourself quite enough for his entertainment.* She wasn't about to be duped yet again into another "hysterical" show of "glandular" behavior over a "petty matter." She smiled airily.

"You couldn't be more right, Mr. Ramsey. Very few men, I'm certain, would have reacted as I did. In the first place, the 'little misunderstanding' about your true identity would never have come up—because if you'd made that cute remark about the view of my . . .

bottom . . . to a *man,* he'd have knocked you into the middle of next week!"

Like a chameleon, Kight Ramsey's expression flowed into one of pure enjoyment, and in spite of herself, Sarah's heart swelled with delight to hear again the deep, rich laugh that rolled out of his chest like spring thunder. "You're a force to be reckoned with, Miss Halston," he said admiringly. "Point and game to you."

"Sarah?" Sarah heard the meek little voice and turned to see Iris's wide-eyed face peeking around the corner from the kitchen. "The oven timer went off."

"I'll be right there, dear," she said, then turned back to the still smiling man. "Well, then, if that's all," she said hesitantly.

He motioned her toward the kitchen. "Don't let the cookies burn. I'll wait, if that's all right?"

Torn between the oven and seeing him out of the house and out of her life, Sarah dithered a moment and then bolted for the kitchen. She quickly divested herself of the ludicrous flippers and the two pillows and made an ineffectual dab at her face with a wet paper towel. "Oh, well, it's hopeless," she sighed, then quickly transferred the cookies from the oven to the counter. Iris pranced about, getting in her way as she put the second batch in. "Now you stay here, love, and wait for the timer again. I'll be back before you can say 'Jack Robinson.'"

As Sarah returned to the living room, where Kight Ramsey waited, she knew she was glad they hadn't parted a few minutes ago, bitterly angry with each other, but still, nothing had changed, really. He'd go now and that would be the end of it. She found him standing near the window wall, looking out into her back garden. He certainly wasn't shy about making himself at home, was he? she thought testily.

"What a wonderfully serene little garden," he said. "Did you by any chance have anything to do with it? Or did it come with the apartment?"

"No, it's my design, with the help of a friend," she said, looking at it afresh, seeing it as he might see it. There were three artfully pruned and trained dwarf Mugho pines in the far left corner, with one small grey boulder off to one side. Clutches of naturalized daffodils, blooming cheerfully in the misty January air, were tucked here and there among the pines. A gentle berm off center in the middle of the area was covered with potentilla, a fine-leafed ground cover that bloomed sporadically throughout the spring and summer with tiny yellow strawberrylike blossoms. On the right side of the garden, to balance the pines, was one glossy camellia bush, heavily covered now with pink flowers. The rest of the area was covered by soft grey bark chips.

"You really are an artist, aren't you?" he said, fixing her with a thoughtful gaze. With a modest shrug, she smiled and murmured, "Very nice of you to say so, at any rate."

With a little shake, as if coming back to the present, he turned from the window. "Well, I've intruded on your afternoon long enough." But as soon as he'd crossed the room and put his long fingers on the doorknob, he turned again, and with that damnably appealing smile on his face, he said, "I suppose even I could whack those roses into shape, couldn't I? Just a matter of chopping off a foot or so of height, isn't it?"

Horrified by the spectacle of carnage that sprang into her mind, Sarah cried, "Don't you dare!" Then remembering that they were his roses, not hers, she back-pedaled. "I mean, I think that would do almost more harm than leaving them as they are. Can't you just hire someone else for the job?"

"It seems I'll have to," he said pointedly. "But it may take a while. I'd already interviewed everyone who answered my ad, and with so many applicants to choose from, I was sure I'd find someone suitable, so I canceled the ad. And of course I was right. I *did* find someone eminently suitable. However, so be it. Now I suppose I'll have to put the ad back in and go through the whole rigamarole again. You can see that it may be several weeks before this problem is off my hands. In the meantime, the roses are languishing from neglect. You pointed that out to me yourself, you'll recall."

Before Sarah could frame a reply, Iris came shyly into the room carrying a plate of more or less mangled cookies. Not presuming to even hold them out to the strange man, she mumbled, "I took them off the sheet myself."

Seeing the child's discomfiture, he stepped forward with a wide friendly smile on his face, and in response to her unexpressed but clearly meant offer, he said, "I certainly would like a cookie. As a matter of fact, I hoped you'd ask."

He took a cookie from the plate and seemed not to notice when it broke in two and one-half fell to the carpet, whereupon Duke, lurking under the sofa, dashed out and fell upon it like a ravenous lion. Mr. Ramsey chewed the cookie thoughtfully, his eyes cast heavenward, in what only Sarah recognized as a mockery of a pretentious wine taster. Enthralled, Iris stared at him with her mouth slightly open. When he'd swallowed, he looked down at her and said in a gravely sincere voice, "Do you know, that was absolutely bar none the best cookie I've ever eaten?"

Iris rewarded him with a beatific smile. "And what's more," he said, taking her hand in his and brushing it with his lips, "I have never met a more charming and beautiful baker." With a delighted sharp giggle, Iris suddenly turned and plunged toward Sarah to hide her face in Sarah's smock. Over her head the two adults laughed briefly at the child's behavior, and then Kight's eyes locked Sarah's in a gaze so penetrating that she felt it as a caress.

"She's very fond of you," he murmured.

"I'm very fond of her, too," Sarah murmured in return.

Sarah became aware of the flush building in her cheeks, and with great difficulty she tore her gaze from his. He blinked rapidly and cleared his throat.

"Yes, well, I've taken up enough of your time. Or have I said that before? I'll be going now. I probably won't get around to fixing those roses for a few days, so if you change your mind, give me a call."

Sarah laughed softly. "As if you really mean that, about chopping at the roses."

"I do indeed mean it . . . in a day or two." He made an abrupt hacking gesture with his right hand. "Off with their heads. Oh, by the way, my phone number is unlisted, but if you should decide to call, you'll find it on your message machine." And with a provocative smile, he was gone.

When the door closed behind him, Sarah turned to

see that Iris' face had the blank expression of a satisfied theater-goer when the lights go on after the curtain. After a moment of pregnant silence, Iris said in an awed voice, "Did you hear what he said about the cookies?"

Sarah smiled, bemused. "Yes. And how about what else he said? Haven't I always told you what a beautiful child you are?"

Iris shuddered slightly, as if throwing off a lovely spell. She turned her clear, blue eyes on Sarah and in a sweetly explanatory tone, she corrected, "No, Sarah, when he said that, he didn't really mean me. He meant you."

Chapter Three

Since she had no intention of working for him, Sarah preferred not to ask herself then just exactly why she so carefully took down Kight's—for she couldn't help but think of him as that now—phone number from her message machine that evening when she'd returned from taking Iris home. The message itself couldn't have been more straightforwardly businesslike. But all the same, even his recorded voice filled her with a new and very pleasant feeling.

"Kight Ramsey speaking, Miss Halston. I have a matter of some urgency to discuss with you. I'll be in Sacramento sometime this afternoon and will take the liberty of stopping by in the hope of finding you in. In case you want to get in touch with me, I can be reached at the acreage." And then he'd given his phone number.

A matter of some urgency. How could finding a gardener be a matter of urgency to a man like him? To someone like Aunt Elaine, yes; Sarah could believe it of a person to whom all personal desires were matters of urgency—but not to an obviously successful and handsome man in his thirties like Kight Ramsey, who undoubtedly had plenty of more interesting fish to fry.

The only urgency about it that Sarah could see, from his point of view, was to get what was for him a tiresome chore over and done with. He'd indicated that he'd found someone eminently suitable for the position. That didn't necessarily mean Sarah. There may have been one or more others to whom he'd offered the position and who'd turned it down for some reason. Sarah hadn't seen the living accommodations, and the salary hadn't been mentioned. A married man, for instance, might have found both inadequate.

Yes, that was obviously the answer. Like most men, he was used to getting what he wanted without a lot of folderol and backtalk. It was "urgent" to him to be rid of the problem so he could get on with whatever it was he did for a living. Well, Sarah wished him the best of luck in solving his problem. From her point of view, the dismal state of the grounds themselves constituted an emergency situation. If weed-killer weren't applied to that lawn soon, the weeds would flower and go to seed, and all too soon there would be no recourse but to disc up the whole area and sterilize the soil before starting an entire new lawn. And those poor roses! In spite of what he'd said, surely he wouldn't just chop them off as if they were roadside weeds, would he?

For the rest of the evening Sarah found that no program on TV or any magazine she had in the house could for long distract her worry-wart mind from that vulnerable rose bed, living with the sword of Damocles over its unkempt head.

And so it was that the next morning Sarah decided to prune the rose bed. But only that; she was as determined as ever not to work for Kight Ramsey on a full-time basis, no matter how "urgent" his need for a gardener. Sarah knew that he'd more or less blackmailed her into her decision, but the fate that awaited the rose bushes disturbed her more than her own discomfiture. The damage to the bushes would be long-lasting, possibly even fatal, but her own feelings would soon be forgotten once she'd put all thought of Kight Ramsey out of her mind.

It took Sarah some minutes to work herself up to dial his number. She wanted to hear his voice, to be in contact with him, and, then again, she didn't. So it was with both relief and disappointment that she was tersely told by Mrs. Mole, who answered the phone, that Mr. Ramsey was out.

"Oh, I see," Sarah said hesitantly, wondering if this was a sign to forget the whole matter.

"Who's calling?" Mrs. Mole demanded.

"This is Sarah Halston. I met you yesterday—"

"As if I could forget," the woman responded sarcastically. "Mr. Ramsey said if you called about the roses to tell you to come at your convenience."

Sarah noted that he'd been sure enough she'd call to leave instructions. In a matching cold and clipped manner, Sarah said, "Thank you. I'll be there this morning." And she hung up before she could be hung up on. It was petty of her, she knew, but really! Where did the woman get off behaving in such an insufferable way to a stranger who'd done her no harm?

A little over an hour later Sarah arrived at the acreage. Pulling on a pair of gauntlets she'd hunted up from the bottom of her toolbox, she got right to work. It was a more pleasant day than yesterday, clear and sunny, the temperature a fresh fifty-five or so. By eleven o'clock she'd finished about half of the bed and looked forward to assuaging the rumbles in her stomach with the lunch she'd packed.

Suddenly the country silence was rent by the sound of a powerful car engine, and Sarah looked up to see a silver Jaguar swoop up the rutted drive and screech to a halt in front of the house. The car had an English drive, so Sarah couldn't clearly see the woman who slid out on the wrong side of the car and went into the house without ringing. Sarah'd got only an impression of blonde, well-dressed elegance. Who was she? She was obviously close to Kight Ramsey, entering the house more like family than company. Perhaps she was family? Sarah thought hopefully. A sister, perhaps. Or a cousin. Sarah was embarrassed at the ferocity of her curiosity and also her transparent wishful thinking. *Face facts,* Sarah told herself, *a man like Kight Ramsey is bound to have a woman like that in his life—someone just as close as family, or closer.* Then, too, there was an even more likely possibility—but Sarah pushed the thought away and bent to her work with a new vigor. After all, she was here on an odd job, and nothing about Mr. Ramsey's situation was any of her business.

It was highly irritating that she persisted in mooning over that exasperating man. And puzzling, too. Sarah'd gone out with a number of men over the years and been fond of some of them; she was presently fond of Bill Blanding, Iris' social worker. But never before had her thoughts revolved so obsessively around a man; never before had she reacted physically in such alarming ways to the touch of a hand, the sound of a voice, the sight of a smile. Really, it was ludicrous! Sarah hadn't thought she was the sort of young woman to go all soft in the head over a man—and a man so far out of her reach, at that. Why, she was behaving like a teen-ager swooning over a rock star! Moreover, Sarah didn't even like Kight's type—the flippant playboy to whom the world was a joke and anything female was fair game.

It was noon when Sarah looked with satisfaction at the mounds of prunings around the bed and went back to her little Datsun to open her packed lunch. She ate her chicken sandwich and was munching the crisp Granny Smith apple when through the windshield of her car she saw the front door of the house open and Mrs. Mole emerge and march toward her with a look of long suffering on her face.

"Mrs. Ramsey wants to see you," she announced.

A terrible ache of disappointment and dismay filled Sarah's throat and chest so that she felt short of breath. How stupid she'd been to think that the lack of a wedding band meant anything in this day and age. Of course he would be married—and to just the sort of woman Sarah'd glimpsed an hour ago. Pride forced her to recover herself in Mrs. Mole's presence. Clearing her throat, she asked briskly, "Now?"

"Yes, now."

Sarah put her apple into her lunchbox, thinking what a wretched sight she must be after a morning's work in the rose bed. But she'd rather die than make the slightest move to tidy herself under the knowing smirk on Mrs. Mole's face. Sarah followed behind the woman into the entry atrium, and during the seconds it took to walk through it, a part of Sarah's mind noticed that the plantings looked dull and feeble, as if they needed a good feeding and some basic grooming. Mrs. Mole preceded Sarah down the hall toward the bathroom she'd used several days ago, and beyond to what seemed a suite of rooms connected to but separated from the main house. Stopping at a closed door, Mrs. Mole knocked, and upon hearing a murmured "Come in," she entered with Sarah reluctantly following.

Sarah's attention was immediately drawn to the two women who sat facing each other on two white loveseats placed on either side of a white brick fireplace. The remains of a tray lunch were on a large cocktail table between them. The young woman whom Sarah had seen arrive a short time ago sat nestled languidly against the soft cushions, her relaxed posture belying the utterly sharp, scrutinizing stare of her chilly blue eyes. The elderly woman sat upright, her demeanor one of dignified confidence. Wings of white hair swept up and back from her strong-boned handsome face, resolving at the back of her head into a traditional French twist.

But even the sight of a heretofore unexpected person in the house seemed less surprising than the unctuous expression that had come over Mrs. Mole's face and the dulcet tone of her voice, so sweet it might have drawn flies, as she said graciously, "Miss Halston, ma'am."

As if this transformation weren't bewildering enough, Sarah was further taken aback when the older woman, not the younger, acknowledged the introduction by reaching forward to shake Sarah's hand. "I'm happy to meet you, Miss Halston," she said with a pleasant smile. "This is Vivica Harrington." She indicated the elegant blonde woman who favored Sarah with a lazy nod. "Mrs. Mole, I think Miss Halston might be glad of a hot drink. Would you prefer tea or coffee, my dear?"

Still casting about for her bearings, Sarah replied, "Oh! Thank you. I'd love a cup of coffee."

"And you, Vivica?"

"Nothing for me, Grace. I really must eat and run, I'm afraid. I'm expecting Kight for drinks later this afternoon, and I have a few errands to run in the meantime."

A flood of relief weakened Sarah's knees. Never mind if Kight was intimate with this woman; at least they weren't married. Not that it concerned her either way, Sarah reminded herself. Vivica Harrington rose gracefully from the couch and placed a light kiss on the older woman's cheek. Turning to Sarah with a smarmy smile, she said throatily, "So fascinating to meet a woman who does physical labor." Her delicate shudder spoke volumes. "But then, to each her own, I suppose."

"I'll see you out," Mrs. Mole said, a triumphant smile playing about the corners of her mouth. Taking up the lunch trays, she followed the model-thin Vivica out of the room and closed the door behind them.

"Please sit down, Miss Halston," Mrs. Ramsey said, gesturing toward the couch Vivica had just vacated. Sarah glanced around the room for a wooden chair that her clothes wouldn't soil, but Mrs. Ramsey saw her intention and added, "Don't worry, my dear, your clothes look perfectly clean, and in any case, the slipcovers are washable. I'm much too practical a woman to have this light color if they weren't."

As Sarah sat down gingerly, her brain stopped

whirling enough for her to wonder for the first time what this summons was about. Perhaps the true purpose would be revealed after Mrs. Mole had brought the coffee.

"No doubt you noticed that the house and grounds are at sixes and sevens," Mrs. Ramsey said lightly. "My son only recently bought the house, and I arrived to take up residence just last weekend. The previous owners had a lingering illness in the family that prevented them from seeing to things as they came up. This was a country house to them, but my son and I intend it to be our principal residence, so, as you see, we have our work cut out for us."

Sarah murmured politely and was relieved from further comment by Mrs. Mole's decorous knock on the door. As she placed the coffee tray on the cocktail table, her attitude toward the older woman was one of devoted deference. She made a point of turning a falsely gracious smile on Sarah, as well, as she virtually bowed herself out of the room again.

"This house has such potential, I think," Mrs. Ramsey said conversationally as she poured Sarah and herself a cup of steaming, fragrant coffee. "I know one is supposed to love and appreciate one's family home, but I must confess I'm glad to be rid of mine. I lived as girl and woman in one of those huge, gloomy old houses near McKinley Park. Perhaps you know the area?"

"Why, yes, I do," Sarah replied, charmed by the coincidence. "I grew up near there, too."

"Really! You'll understand, then, how an old woman living alone in one of those mausoleums would prefer a cozy relaxed house like this. And I'm very grateful to

Kight for insisting that I join him here. It's not every bachelor son who would want his creaky old mother underfoot."

Sarah demurred that Mrs. Ramsey was a creaky old woman, but privately she agreed that, creaky or not, a live-in mother would not be welcome to many bachelor sons. It showed a side of the man that Sarah wouldn't have credited. She had the impression that there were funds enough in the family to settle the mother in comfortable private quarters of her own if size and convenience were the only desiderata.

"Of course, I like to think my presence will be useful to him, too," the charming older woman went on. "Because of his business, he keeps a condominium in Sacramento and an apartment in San Francisco, but now that he's passed thirty he seems to feel the need for a real home, for an emotional base, don't you know. And until he marries and has a wife to make a center for his life, I hope to do that as much as I'm able."

"I see," Sarah said. Perhaps this lady had something to do with her son's urgent need to get the grounds in order. Sarah felt sure that the gardens of Mrs. Ramsey's previous house were as well kept as her Aunt Elaine's had been. She might appreciate the informality of this rambling house, but she wouldn't care to live in a wilderness. And surely the fastidious Vivica, whom Sarah felt with heavy certainty was the prime candidate for the future center of Kight's life, would expect a setting that did justice to her beauty. Sarah's wandering attention veered back to Mrs. Ramsey as she heard her say, "I see by your résumé that Ben Yashimoto trained you. In Sacramento, no one could boast more sterling credentials than that."

"Oh, you know of Ben?" Sarah asked, gratified to know that dear Ben's name and fame lived after him.

"Indeed, yes. Elaine Lang's garden was the envy of all Sacramento. People offered him the most outrageous bribes to leave her and work for them, but he never even seemed tempted. No one could understand why, since there were many who felt that Elaine didn't properly appreciate him, and to work for almost anyone else would've bettered his position."

Sarah felt the color rise in her cheeks. Although much of what Mrs. Ramsey was saying was news to her, and most fascinating, Sarah knew she must speak immediately before the older woman said something about Aunt Elaine that both of them might find hideously embarrassing.

In a reminiscent tone, Mrs. Ramsey began, "I remember one garden party Elaine had—"

Sarah burst out, "Elaine Lang was my aunt, Mrs. Ramsey. She raised me. That's how I came to be trained by Ben, you see." She maneuvered a social smile onto her lips so that Mrs. Ramsey would feel that no harm had been done. But the woman had the blank look of shock on her face. "Oh, my dear girl, what have I said! Have I said anything I shouldn't have?" she asked with a worried, distracted air.

In a soothing tone, Sarah hastened to reassure her. "No, of course you haven't. I, too, know that Aunt Elaine didn't appreciate what a jewel she had in Ben."

But Mrs. Ramsey seemed devastated, perhaps by her private thoughts of what she might have said if Sarah hadn't stopped her. Because she still looked so distraught, Sarah's soft heart yearned to allay her feelings. Picking her words carefully so as not to be disloyal to her own kin, she told Mrs. Ramsey in a light, amused voice, "You know, Ben explained to Aunt Elaine over and over that a rose blossom should be cut just above a five-leafed node, but she *would* persist in cutting the flowers willy-nilly so that the stems suited the vase she had in mind. It got so he and I conspired to prevent her from ever cutting her own bouquets."

Sarah was relieved when Mrs. Ramsey's face relaxed into laughter. While she refilled the coffee cups, the woman's expression became thoughtful. "So you were the little niece Elaine took in after her sister's death." Then, seeming to shake off shadowy memories of the past, she looked up and smiled impishly, and Sarah's heart recognized the origin of Kight's appealing smile.

"Do you know, I believe we've met, although you were much too young to remember. I paid a call on Elaine one day, and during the visit I mentioned that a mutual friend of ours had praised a magnificent Peace bush of hers. I asked if I might see it and she took me out to the garden. I noticed Ben digging in what I think must have been a perennial bed way across the lawn. A little girl was playing nearby him. When Elaine and I reached the rose bed, she looked at all the bushes in a baffled way and I could tell she hadn't a clue as to which was the Peace bush. Then she reached out toward a white rose and suddenly Ben called out and came loping across the lawn with you fast on his heels. Puffing from the run, he smiled and bowed and said that Miss Sarah would be pleased to cut any rose we wished. Elaine said, "Mrs. Ramsey would like a Peace rose, Sarah." And even though you couldn't have been more than eight, you took the clippers from Ben and without a second's hesitation went to the Peace bush and snipped off three buds at just the perfect cutting stage. Do you remember that?"

Sarah smiled with a nostalgic pang in her breast. "No, I'm sorry I don't remember the particular incident, but that sort of thing happened often. And funnier still, I really don't think Aunt Elaine ever realized what was going on."

Mrs. Ramsey murmured something which sounded like "No, she wouldn't," but Sarah thought it better not to pursue it.

Declining another cup of coffee, Sarah said, "This has been most pleasant, Mrs. Ramsey, but I think I'd best get back to your rose bushes if I'm to finish today."

"Of course. I'm being a selfish, garrulous old woman. But before you go, I wonder if you'd listen to my tale of woe, Sarah. Do you mind if I call you Sarah?"

"Of course not. Please do."

"Thank you, dear. The thing is, I've rashly promised an organization I belong to that they could use these grounds for a summer charity affair. The details aren't settled yet, but we were hoping for a date in June. Kight has been a dear, trying to engage a gardener for a permanent position, but as you know"—here her eyes began to twinkle—"he hasn't been successful." Mrs. Ramsey paused and a bemused expression came over her face. "Kight has always had a volatile nature, but I must say I've seldom seen him so disgusted and put out by a straightforward task such as this. Shortly after I arrived Sunday afternoon, he came in looking as if he could chew nails. He said one diamond had surfaced among the dross but that one refused the job. 'I've done all I can.' he said. 'The rest is up to you.' When I pressed him, he reluctantly told me what had happened the day you came to interview."

Sarah felt terribly embarrassed that Mrs. Ramsey knew the details of that wretched incident. She'd have given a lot to have been a mouse in the corner hearing just how he'd explained his brash behavior to his mother. With a dismissing gesture, Sarah tried, for his mother's sake, to make light of it. "It was just a silly misunderstanding, Mrs. Ramsey. I'm afraid I was much too touchy."

"Perhaps, and perhaps not, dear. Kight does have a tendency to take liberties sometimes. But, in any case, the upshot of it all is this: I've agreed to take over the project of engaging a gardener since I'm the one in such a hurry. Therefore, the gardener will be my employee, not Kight's. Would that make any difference to whether or not you'd accept the position?"

Sarah hesitated, thinking. How much would this change of circumstances affect her? There was no question that she wanted the job itself. And surely it would be heaven to work for a woman who actually knew there *was* a perfect stage of bloom during which a rose should be cut. And if she needn't be accountable to that man . . . "Yes, Mrs. Ramsey, I think it would make a difference to me," she said with a smile.

Mrs. Ramsey smiled with relief. "Oh, I'm so glad, Sarah. We'll have such a good time turning this place into what it should be. And I want you to know that I have the deepest admiration for your choice of calling."

Sarah ducked her head to hide the sudden tears of gratitude that stung her eyes. Aside from Ben, Mrs. Ramsey was the first person ever to praise her for choosing to be a gardener.

Lowering her voice confidentially, Mrs. Ramsey added, "Aren't you thankful that you're young today, when a girl can do what interests her? Do you know, if I were a girl today I'd like to be an airplane pilot."

Sarah laughed ruefully. "I just wish more people saw things the way you do, Mrs. Ramsey."

"Oh, piffle to them, my dear. They're just small-minded people with no imagination, afraid if they get out of their ruts they'll lose their way. You must learn to pay them no attention."

Sarah felt quite euphoric as Mrs. Ramsey threw on a warm coat and prepared to show her the living quarters. The salary they'd settled on was more than generous; Sarah would more easily be able to help the Reillys over the tight spots now. As they passed through the large living room opposite the hall from the atrium, Mrs. Ramsey said that the next cold or wet day Sarah could begin the rejuvenation of the neglected tropical plants. As they crossed the brick terrace and walked toward the poolhouse, Sarah said, "I have a cat, Mrs. Ramsey. I hope that's all right."

"Perfectly all right. I'm very fond of cats," she replied, striding along briskly, as if enjoying the fresh country air. Nothing creaky about this lady! Sarah thought admiringly.

"What's your cat's name?"

"Duke," Sarah replied.

"For John Wayne?" Mrs. Ramsey asked with a smile.

"Exactly!" Sarah felt a sudden camaraderie so strong that she nearly hugged Mrs. Ramsey. How *en rapport* she was! "I named him Duke because he's tough on the outside and a softy on the inside."

Laughing together, they reached the poolhouse and

Mrs. Ramsey unlocked the door and pushed it open, gesturing for Sarah to enter. The little house was approximately the same size as Sarah's present apartment, but much more interestingly designed. It rambled like the main house, but on a proportionately smaller scale. The main room was large, with a white brick fireplace and a dark slate hearth. The walls were creamy white and the carpet a soft pussy-willow grey. The windows had wide sills that cried out for potted plants, and tailored white woven wood blinds had been hung within the frames to clear the sills.

"Come see the kitchen," Mrs. Ramsey invited.

It was the most cunning kitchen Sarah'd ever seen, complete with every appliance ever invented. The cupboards, a sunny yellow Formica that perfectly matched the ample countertops, were obviously new. The appliances were brushed stainless-steel and their soft grayness added just the proper note of seriousness to the gay and cheery yellow in the room.

"Do you enjoy cooking?" Mrs. Ramsey asked.

"Yes, and I'm quite good at it, if I do say so myself."

"Really! Perhaps you'll invite me for lunch one day," the older woman said forthrightly.

Sarah murmured that she'd be honored if Mrs. Ramsey would come to lunch someday, but she wondered if there'd be difficulty in maintaining an employer-employee relationship with such a friendly, democratic person as Mrs. Ramsey. Changing the subject, Sarah said, "Since this is the poolhouse, won't it be inconvenient for your guests having me here when summer comes?"

"Oh, no. The guest rooms are on the other side, nearest the house. What's more to the point, I hope the guests don't inconvenience you! However, we'll do our entertaining on the weekends, and I expect a lovely girl like you will have better things to do with her free time than stay at home on weekends." Then, seeing the dubious look on Sarah's face, she added vehemently, "But you must never feel you have to leave your home if you don't care to. And I imagine you'll want to have your own friends here on weekends from time to time. I know it will all work out famously. The Ramseys aren't a particularly social family, anyway."

Perhaps not, Sarah thought privately, but she'd bet dollars to doughnuts that all that would change when the chic Vivica supplanted Mrs. Ramsey as lady of the house.

When all the immediate details had been settled—such as when Sarah would move in, what her working budget for extra help and supplies would be, and so forth—she left Mrs. Ramsey with a firm handshake and thanks on both sides and went back to finish pruning the rose bed. As she worked Sarah thought about all that had transpired. She felt that it was a day she'd remember always as one of the happiest of her twenty-four years. She had a domain of her own now, something that in a very real way belonged to her, as the land does belong to those who work it. If she never had more than this in her life, if she never had her own family, she would nevertheless be content.

It was four o'clock and already growing dark before Sarah had stowed all the prunings into black plastic bags for the trash men to collect. Eventually she'd establish a compost pile, but in any garden there were always materials, rose prunings among them, that were too woody or thorny to break down easily into compost.

Sarah gathered up her tools and was opening the hatchback of her Datsun when, seemingly out of nowhere, the black Porsche came skimming over the potholes in the drive (she'd have to have them seen to) and glided to a stop just ahead of her car. Perhaps because she was in such a contented mood and feeling at peace with the world at large, every fiber of her being responded with pleasure at the mere sight of Kight Ramsey. Nothing that had happened in the past seemed of any importance at just this moment. Maybe it wouldn't last—nor should it last—but right now she was delighted to see him, so vital and full of masculine energy as he gracefully climbed out of his low-slung car and came toward her with that irresistible impish smile on his handsome face.

"What a sight for sore eyes you are," he said by way of greeting, "although you also look very fetching with red spots and a purple mouth!"

Laughing, Sarah said, "You don't seem at all surprised to see me."

"I knew Mother would pull it off," he said grandly. "She's never settled for anything but the best—and I've learned it from her."

"But what made you think I'd come at all?"

"Oh, I know you artist types—ready to walk through the fires of hell for the sake of your work. I took a chance that the risk of running into me again was less dreadful to you than the threat to the roses," he teased. "And I knew if Mother met you she'd move heaven and earth to convince you to stay."

"Well," Sarah began, feeling abashed and at a loss for words, "news certainly travels fast."

"Good news, yes. I called Mother to find out if the coast was clear. I couldn't be absolutely certain you'd succumb even to my charming mother, and if you hadn't, I wasn't up to facing your wrath so soon again."

"You make me sound like a virago!" Sarah laughingly protested.

"No, never that. But you are a worthy opponent."

In the little silence that ensued, it seemed to Sarah that he could say anything he pleased now and she'd think it clever and funny. "At least now I won't have to starve to death," she rejoined, and she was rewarded when he threw his head back and laughed in that way that did such alarming things to her heartbeat. The smile lingered on his face as he looked searchingly at her in the growing dusk.

"You look exactly like a fresh red rose—do you know that?" he asked softly.

Sarah blushed and turned her face away.

"No, don't look away," he said, stepping nearer to take a lock of her hair in his fingers. "I think I'll call you Red, if you have no serious objection."

Sarah hadn't the strength to make a reply; all she could do was stand transfixed, her gaze locked to his.

"I see you remember that you said you'd rather starve than work for me," Kight said dryly, "but now that you're safely working for my mother, may I tell you something? I, too, would rather starve than have you work for me."

Sarah's lips parted in a gasp of surprise at such an outrageous remark, but before she could gather her wits to reply, his mouth came down on hers with the suddenness of a hawk diving for its prey and his arms wrapped around her with such strength that, struggle as she might, she was held fast. As the pressure of his lips increased on hers, she felt a warm tingle steal through her body, and her mind filled with a languorous mist. There was soon nothing in the world but the feel of his mouth on hers and his hard, lean body pressed against the whole length of hers, soft and yielding now in a way it had never been before.

Suddenly a voice with the stridency of a sea gull's cry called out his name. "Mr. Ramsey! Important telephone call for you."

Snapped back to her senses, Sarah jerked away from him to see Mrs. Mole standing in the open door of the house. It was too dark to make out the expression on her face, but it was light enough to see the shocked disapproval in her rigid stance. Now the awful woman would add this to her already full bag of resentment and dislike of Sarah. Shaking herself like a ruffled pigeon, Sarah turned on Kight with blazing indignation.

"Just what was the meaning of that last remark? And what gave you the right to . . . to behave in that . . . that . . ." she sputtered.

"The right of a man with a maid, my lovely red rose," he drawled. "And as for my last remark, if you're not busy tonight, give it some thought. I'm sure a smart girl like you can work it out."

"Beast!" Sarah shouted after him as he turned to lope toward the house. His only reply was to turn and wave nonchalantly before he broke into a run to answer his damnable important phone call—and it didn't take much figuring out who that was from! Sarah bit down hard on her lower lip to stop the tears of anger that swam in her eyes. She'd be boiled in oil before she ever let him take advantage of her again! *The right of a man with a maid—indeed!* she thought venomously.

Sarah stamped on the accelerator of her modest little car and shot away down the rutted drive and toward the home she'd just abandoned for heaven only knew what miseries.

Chapter Four

Fuming, Sarah pushed her little car to its utmost in her desire to put as much distance as possible between herself and that insolent and arrogant man. Even a halfwit wouldn't need all evening to work out the meaning of that last outrageous remark of his—that he'd rather starve than have her work for him. He was obviously the all-too-typical man who was afraid he'd be laughed at for hiring a woman to do "a man's work." It was just peachy for him to have "the best" gardener to work on his ratty grounds, even if the best was a woman; but far be it from him to take the responsibility for it. Leave that to his mother!

And then, too, if Sarah was an employee of his, he wouldn't be free, as he obviously thought he was now, to foist his unwanted advances on her. Oh, Sarah knew that many men employers did take advantage of their female employees in that way, but somehow, as much as she detested him, she felt sure Kight wasn't the kind of man to take advantage of someone dependent on him for her livelihood. Instead—give the devil his due!—he'd fixed it so he could have his cake and eat it, too. Well, he'd picked the wrong pigeon this time!

Sarah'd agreed to take the job, and she wanted the job. She wasn't going to let him spoil everything, either for herself or for his mother. But she'd make sure he understood the ground rules: Sarah was Mrs. Ramsey's employee, not Kight's; and she wasn't a girl who indulged in casual flirtations. Let him spend his kisses and compliments on the beauteous Vivica.

When Sarah arrived home there was a message on her answering machine from Bill Blanding. When she returned his call, he invited her out to dinner.

"I'd like nothing better," she said, pleased. "I feel like celebrating tonight. Remember that job I applied for? Well, I was hired."

"Terrific! In that case I'll take you to Old Sacramento. This calls for something special," Bill said fondly. "It isn't every day that a girl's dream comes true."

No one could ask for a more loyal friend than Bill, Sarah thought as they said good-bye and she hung up the phone. She'd take more care than usual with her dress this evening to do him proud. But first, she wanted to speak to the manager of the apartment complex about moving out.

Mr. Nelson seemed to be expecting her from the impatient way he nodded as she began to tell him the reason for her visit. "Yes, yes, I know," he said. "Mrs. Ramsey called just minutes ago and explained the whole thing to me. Naturally, I'll be happy to do as she asks. The last thing I'd want to do is stand in your way, Sarah. This is a wonderful opportunity for you," he said in an awed voice. "With the Ramseys as future references, you can pretty much write your own ticket."

"Whoa!" Sarah halted him. "Instead of my telling

you my plans, maybe you'd better tell *me,*" she joked. "What did Mrs. Ramsey ask that you'll be so glad to do?"

"She said the two of you agreed that you'd start work there right away, but that you wouldn't move up there until you'd lived out your thirty days' notice here."

"That's right," Sarah said. "I have my maintenance agreement with you, Mr. Nelson. I wouldn't leave you in the lurch."

"Never mind that. There's very little to do in January, anyway. Mrs. Ramsey said if it was all right with you and me, she'd pay your last month's rent in lieu of the thirty days' notice so you could get settled sooner and avoid that twenty-mile drive twice a day. I think it's very handsome of her, don't you? Not that one would expect less from the Ramseys."

Sarah saw that Mr. Nelson knew something about the family that she didn't know, but rather than be laughed at for ignorance again, she nodded, murmuring her agreement. When she and Mr. Nelson agreed on the final details she returned to her apartment to dress for dinner with Bill.

After a shower and shampoo had washed away the day's work grime, Sarah felt much better. Wrapped in a white terry-cloth robe, she blew her auburn hair dry as Duke, in a rare kittenish mood, sat on the bathroom counter and batted at the swinging cord of the hair dryer.

"Wait until you see your new home, Duke. Five acres for you to patrol, and a lovely fireplace to lie in front of for your evening snoozes."

When Sarah's hair was dry and the ends were curled under, she applied a light coat of foundation to her magnolia-colored skin and a smidgen of blusher to accent her high cheekbones. A subtle application of sea-foam eye shadow and a few swipes of mascara on her thick brown eyelashes brought to her face a more dramatic glow that she, in her unawareness of her own beauty, realized.

From her small selection of dressy clothes, she chose a jade-green jersey dinner dress with long sleeves and a cowl neck. The clinging material molded softly over her ample bosom and fell gracefully from her shapely hips. She chose a pair of high-heeled sandals in grey lizard to complement the dress. She was putting on the pair of jade earrings that Aunt Elaine had sent for her last birthday when the doorbell rang. After a quick application of a musky perfume behind each ear, Sarah ran to open the door.

Bill Blanding looked as if he, too, had taken special care for their evening out. He wore a pair of dark brown slacks and a sport coat Sarah didn't remember seeing before in a brown and mustard herringbone pattern. A dark brown tie set off his curry-colored shirt. His dark gold curly hair seemed freshly cut and styled, and the admiring smile on his pleasant, open face was reflected as well in his warm, brown eyes.

During the short drive to Old Sacramento, Sarah filled Bill in on her new position, describing the estate and its whereabouts, her new quarters in the poolhouse, her duties, and so forth. As she chatted on, Sarah was surprised at herself for concealing from Bill what took place during the first interview with Kight. A week ago she'd have related such a story to Bill as an amusing joke on herself and they'd both have laughed at it. But now she felt she didn't want to share with anyone, not even dear Bill, anything at all about Kight Ramsey. Well, there was an obvious explanation for her reticence—she was still smarting from the several humiliations Kight had subjected her to, and she didn't want to appear the dolt in Bill's eyes.

Bill parked his car in the area's central parking garage, an old renovated brick warehouse, and they set out to walk the few blocks to the restaurant.

"It sounds like an ideal job for you," Bill said, giving the arm she'd tucked through his a friendly squeeze.

"Oh, it is, Bill. And my boss is such a nice woman, and knowledgeable about gardens herself. It'll be a pleasure to work for her."

Past six o'clock now, the winter darkness had fallen. The brick streets, moist from the misty air, gleamed romantically in the circles of light that drifted down from the beautifully restored iron and leaded-glass streetlamps. On this Monday evening in winter the area was less crowded with tourists and citizens than Sarah'd ever seen it, adding to the illusion that they were back in time some hundred and twenty years, walking down the street of the roaring young frontier town that sprang up virtually overnight when the hordes poured into the Great Central Valley in search of gold.

Sarah enjoyed the sound of their footsteps on the raised wooden sidewalks as they passed by boutiques, bars, restaurants, and specialty shops, all giving new life to the old brick and wood flat-faced buildings, restored and rebuilt to look just as they had in the 1850s.

"Here we are," Bill said, guiding Sarah into the

doorway of a building that had once been a thriving bank in those exciting days of boom and bust. Inside the tiny foyer they waited for the open brass-barred cage of the old elevator to slowly creak its way down to them from the floor above, where the dining rooms were. To their left was a wide stairway deeply covered in forest-green carpeting.

"If you don't mind, we could walk up faster," Bill suggested.

"Let's do," Sarah agreed. "Frankly, even if they've put in all new machinery, I still don't trust that old thing," she said with a laugh.

The broad steps and the solid oak handrail, its color deepened to a golden patina from thousands of hands over the years, made the climb an easy one. Bill stopped at the reception desk, once a teller's cage, and gave his name to the reservations clerk. Then they went into the bar to wait for their table and Bill ordered himself a martini on the rocks and a glass of white wine for Sarah.

Making light conversation while they waited for their drinks, Sarah said, "I do love to come here and see the old buildings and the sidewalks and all the careful craftsmanship, like those Doric ceiling moldings and the mahogany wainscoting on the walls—" She stopped and smiled ruefully. "Can't you imagine how pathetic I'd sound to a European, carrying on about a building that's only a little over a hundred years old?"

"Never mind." Bill laughed. "You live in a young country. Even age is young here."

"Just look at that, for instance," Sarah continued, indicating the long mirror behind the bar. "Where would you ever find such beautiful etched glass today? Why, it looks more like lace than glass."

Their drinks were set in front of them just then, and Sarah raised hers to toast Bill's and her own reflection in the mirror. With a start so severe that her heart missed a beat, she saw the dark, sardonically smiling face of Kight Ramsey suddenly appear over her own right shoulder, and beside him an annoyed-looking Vivica Harrington.

"Good evening, Miss Halston. How pleasant to see you again so soon," Kight said in the dry tone that often signals a private joke. Sarah swallowed hard and nodded in greeting to his reflection. Then, coming to her wits, she spun around to face him directly. From the corner of her eye she was aware of Bill's curious expression. Sarah introduced them to each other, privately marveling that she could remember her own name, let alone any of theirs, so giddy did she feel. Bill offered his hand to them both, and he said to Kight, "I'm very pleased to meet you, Mr. Ramsey. Sarah was just admiring the work you did down here."

What on earth was Bill talking about? Sarah wondered, and why the note of deference in his manner. Bill went on to say, "Did you by any chance have anything to do with this particular building?"

"I'm happy to say I did," Kight replied, responding as everyone did to Bill's warm friendliness. "And we did the parking garage and some of the buildings on Second Street."

"You did a fabulous job. Sarah and I were just remarking that it's like being a hundred years in the past."

"Very nice of you to say so, Mr. Blanding. We all enjoyed working on the project. It was a nice change from the usual dams and bridges and office buildings," Kight said.

As the two men chatted amiably, Sarah had a chance to notice what Vivica Harrington was wearing. Her blonde hair was coiled into a Psyche's knot at the back of her head, revealing the good bones in her thin face. Her dress was a heavenly French blue; its sheer crepe blouson bodice, held up by the merest suggestion of spaghetti-thin straps, showed off the milky-blue alabaster color of her shoulders. The staggered hemline of the dress drifted delicately with every move of her small silver-slippered feet. All in all, Sarah sighed mentally, a girl impossible to compete with.

Vivica, growing bored with the men's attention being wasted on each other, broke in to say to Sarah, "What a pretty little dress you're wearing, Miss Halston. You look so different from when I saw you earlier today that I might not have recognized you if Kight hadn't pointed you out."

Sarah's temper flared, but just in time she remembered what Ben Yashimoto had taught her of the Oriental martial arts: an opponent's own aggression turned back upon himself is always more effective than your own aggression turned on him. In a sweetly docile tone, Sarah replied, "You, on the other hand, looked just as beautiful earlier as you do tonight, Miss Harrington."

Vivica flushed an unattractive red as she realized that her cattiness had been fully exposed by Sarah's flattering response. The two men looked down at the floor with surreptitious grins on their faces, and Sarah had the further reward of knowing that she'd spoken nothing but the truth.

Sarah was relieved to see the waiter appear to lead them to their table before Vivica had a chance to recover and retaliate. She and Bill turned over their seats at the bar to Kight and Vivica and followed the waiter into the dining room. When they were seated, they were given their napkins and flatware in authentic old iron safety deposit boxes, dented and rusted from hard use and age. Spreading his napkin on his lap, Bill said, "Do you suppose these boxes once held gold nuggets and deeds to mining claims?"

Sarah, still thinking of the recent meeting, nodded absentmindedly. Bill continued: "Speaking of success and riches, how do you happen to know Kight Ramsey?"

"He's the son of the woman I'm going to work for," Sarah explained. "How do you happen to know him?"

Bill's laugh was explosive and incredulous. "Don't you ever read the newspapers—not even the society columns? Why, everyone in California knows who Kight Ramsey is. Ramsey Construction is one of the biggest heavy-construction firms in the United States; they build all over the world."

Sarah was nonplussed. "Is he *that* Ramsey? Of course I've heard of the company—I've seen Ramsey signs at construction sites in the city all my life—but, no, I *don't* read the society columns, and I've never heard the first name of the owner before, so I just didn't connect the name. Well, what do you know about that?" she mused. "No wonder Mr. Nelson was all agog." So that's why Kight thought he could ride roughshod over the feelings of others to get what he wanted, Sarah thought. He was the head of a tough, competitive business where success didn't come from being a shrinking violet.

Bill went on to say that Kight's father started the business with a G.I. loan after World War II, and, with the help of his partner, George Harrington, built it into one of the largest companies in California. "Since his father's death about ten years ago, Kight has expanded it to an international concern. He's quite a guy," Bill said admiringly.

"George Harrington?" Sarah repeated. "I wonder if that's Vivica's father." If so, that meant they had a large business in common, Sarah thought, depressed; even more reason for them to marry. As if her thoughts were a magnet, she saw the two of them enter the dining room and follow the waiter to an intimate table for two in a cozy corner of the mahogany-paneled dining room. With a heavy, burning feeling in her chest, Sarah saw Vivica reach across the table and touch Kight's hand, then smile fondly at him as she spoke words Sarah couldn't make out.

When the waiter came to take their order, Sarah was glad to let Bill take charge. She doubted that she would have much appetite, anyway, so tense and fluttery did her stomach feel. Bill ordered a fresh asparagus bisque for the first course, followed by rack of lamb with new potatoes, and a butter-lettuce salad. "We'll decide on dessert later," Bill told the waiter.

During the quietly efficient service of the dinner, Bill related to Sarah the details of the wearing day he'd had. With a caseload of some twenty families, he lived daily with an incredibly heavy emotional burden. When

Sarah reflected the depth of emotional involvement required just by her own commitment to Iris and the Reillys, she marveled that Bill could bear up under the strain of twenty families' problems. Sarah often thought that Bill's dedication to his work kept him from making any deep personal commitments in his private life. As far as she knew, there was no woman who was more than a close friend to him, as Sarah was herself. It was as if he spent his entire supply of love on his clients, saving none for his private life.

And just as Bill's clients' lives were his main interest, so were they his main topic of conversation. Over the past year that she and Bill had become friends, Sarah'd grown to know and sympathize with his flock. It was something like following a soap opera on television, but with the very grim difference that the miseries were real, the triumphs few, and one couldn't switch the button to Off and forget the whole thing.

The interest Sarah usually felt in Bill's work was simply not there tonight, she thought with chagrin. Again and again her attention lapsed as her eyes, as if by their own volition, strayed to the pair in the corner of the room. They, too, had been served their dinner now, and Sarah was much more fascinated by the sight of Kight's expression as he ate his food while listening to Vivica talk. It wasn't until Sarah heard Iris' name mentioned that her wandering attention snapped back to what Bill was saying.

"Iris is doing very well. I just wish I could say the same for her mother."

"You've been to see her?" Sarah asked eagerly.

"I went to the hospital yesterday."

"And there's no change?"

With slumped shoulders, Bill stared down into his dinner plate and pushed a chop bone with his fork. "A change for the worse, if anything. She didn't recognize me."

"Oh, no!" Sarah exclaimed. "I thought . . . since she came out of the coma—"

"The hospital staff says she doesn't recognize anyone. She just lies there in her bed, doesn't move, doesn't speak. The doctors say there's nothing they can do; the damage to her brain was too severe. Soon they'll move her to a nursing facility and that'll be the end of it."

Sarah's body grew cold with despair, and in a bitter voice she said, "Jack Millidge has taken his wife's life from her just as surely as if he'd killed her. And look what he's done to Iris' life, as well! Not that she isn't better off without him."

Bill nodded in agreement. "The question is: How long can we count on him staying out of her life?"

"Yes," Sarah agreed. "As things are, he could drift back into town any day and find out where she's living. I just thank God the poor little kid doesn't realize she's living with a time bomb. I'll never understand why the district attorney didn't prosecute Millidge," Sarah said hopelessly.

"Well, you know what he said," Bill replied. "The courts don't like to put men behind bars for what they still think of as 'private domestic matters.' And to be fair, at the time no one knew how badly Mrs. Millidge had been hurt. The doctors expected that when she came out of coma she'd eventually recover."

"Yes, until the next time!" Sarah said heatedly.

Bill shrugged in resignation. "That's the way it is, Sarah. All we can do is work within the law."

"Even when the law is blatantly wrong?" she demanded.

Bill reached across the table and covered her clenched hand with his. "Sarah, when the law is wrong—and I admit it sometimes is—we must work to change it, not break it."

"I suppose so," Sarah muttered begrudgingly. Bill was right in theory, of course. But in practice, was it right to gamble with a child's very life in order to obey a law based on outmoded social tradition? Bill was the best of men, sometimes nearly saintly. But he was too mild natured, respectful of authority to a fault; he'd never come out on top in a real knock-down, drag-out fight. Privately, Sarah vowed that if Iris' father ever again threatened the child's safety, she'd do everything in her power to thwart him—law or no law.

"Let's not talk about it anymore," Bill said, patting Sarah's hand. "I'm sorry. I was wrong to even bring it up on an evening that's supposed to be a celebration." He tried to grin but produced only a weak, sad smile. Sarah nodded faintly. They could change the subject, of course, but she knew the evening was ruined for them both. Sarah didn't even want to enjoy herself now, not with Mrs. Millidge lying motionless in a hospital and little Iris' future shadowed with the threat of yet more violence and neglect.

"They have wonderful desserts here," Bill said in a determinedly bright tone. "Let's both splurge and have the sweetest, gooiest concoction on the menu."

Sarah forced a smile. "All right. Why not?" She felt a sudden rush of affection for Bill and a pang of shame at her recent thought that he'd prove ineffectual in a fight. At least no one could ever say that he didn't have Iris' best interests at heart just as much as Sarah did—and twenty other families' as well. Her smile was warm as she reached across the table to touch his cheek in a gesture of loving friendship.

As Bill gave their dessert order to the waiter—"devil's food cake with fudge icing for me, and a butter pecan sundae for the lady"—Sarah saw Kight and Vivica rise from their table and make their way not toward the foyer, but directly to Bill's and Sarah's table, arriving just in time to overhear the waiter repeat the order for what suddenly struck Sarah as embarrassingly childish food for two adults to eat.

Sarah was sure that the smirk on Vivica's face proved she thought so, too. Bill made a motion to rise, but Kight motioned him to keep his seat. "We're going to a new disco club down the street," Kight said in an oddly stiff manner. "Would you like to join us?"

"Oh, please do," Vivica said vivaciously. "We're meeting a group of friends there. It should be such fun."

Bill glanced at Sarah and read the "no" in her eyes. "I'm sorry, we've just ordered dessert and we'll be here a while. Thanks so much, anyway," Bill replied.

Vivica exclaimed, "Just cancel your order! Refined sugar isn't good for you, anyway. I never eat sweets," she said. Sarah could have sworn that Vivica moved her body slightly to draw attention to its remarkable slenderness.

Kight dryly interjected, "I think these people can

decide on their own diets, Viv. We'll be going along, then."

But Vivica, not one to take no for an answer, went on to say, "Well, then, finish your meal and then come along. Please say you will," she coaxed.

Sarah could see by the pained look on Kight's face that it wasn't he who was pining for their company. No, it was Vivica. But what possible reason could she have?

"I'm sorry, no," Sarah said firmly. "I've had a very tiring day, and so has Bill. I'm afraid disco dancing has little appeal for either of us tonight."

"Of course," Vivica cooed. "You must forgive me, Miss Halston. I forgot for a moment that your work entails physical labor. But I must say I'm devastated. Our friends would've been so amused to meet a lady gardener."

A look of shocked disapproval came over Kight's face, and Sarah realized that he hadn't thought Vivica would be so candid about her reasons for inviting Sarah to join their friends. His own reluctant invitation indicated that he'd find it distasteful, not amusing, to introduce a "lady gardener" to their elegant friends.

Sarah felt an indignation that was almost exaltation building up in her, and as it grew she forgot everything Ben Yashimoto had ever taught her about passive aggression. She might not be able to strike back at Jack Millidge, nor at the unfairness of life itself—but she could certainly put this insulting and catty woman in her place, and Kight Ramsey, too, once and for all!

With a cool, dignified smile, Sarah replied, "I'm sorry to disappoint you, Miss Harrington, but since your friends are so easily amused, they might even prefer my clown act. Children of all ages find it very entertaining. Mr. Ramsey saw it at my apartment yesterday afternoon. He'll vouch that it's amusing, I'm sure."

No dessert in the world could have compared with the sweetness of Sarah's revenge when she saw the venomous glare that Vivica Harrington, on hearing Sarah's reply, directed at Kight Ramsey.

Chapter Five

During the next week Sarah finished all the previously contracted gardening jobs for her old clients, all of whom were very sorry to lose her. The evenings were spent packing her belongings in food cartons cadged from the nearby supermarket. Duke was much in evidence, weaving in and out between Sarah's legs as she worked, investigating each empty box for possibly tasty insect life, and meowing conversationally about this interesting domestic upheaval. Mrs. Ramsey wasn't expecting Sarah to begin work until the following Monday, and Sarah welcomed the interim days when she could rely on not running into Kight.

On the way home from their dinner date last Monday night, Bill had expressed amazement at the way Sarah had lashed out at Vivica Harrington. "I've never seen you blast anyone like that. It showed me a whole new side of your character," he said.

"Well, she deserved it!" Sarah said forcefully. "That was only the second time in my life I'd met her, and she'd already insulted me four times!"

"But why should you care so much what she thinks of you?" Bill asked, puzzled.

"I don't care what she thinks! It's just very annoying when someone her age, and a woman at that, can't get it through her head that work has no gender. Only living things have gender. And besides, why should I stand by meekly and let anyone insult me?" Sarah exclaimed.

"Well, I guess you're right," Bill said doubtfully. "But from what I've gathered in the past, you're usually a lot more tolerant of that kind of attitude toward your work. If you don't mind my saying so, I wondered if you mightn't be jealous."

"I am *not* jealous!" Sarah cried. "What a ridiculous idea!"

Bill stole a sidelong glance at her furious face and said mildly, "That's good, because I'd hate to see any friend of mine set her heart on Kight Ramsey. According to the gossip columns, he's not the marrying kind."

"You can set your mind at rest about my heart, Bill," Sarah said grumpily. "I couldn't care less if Kight Ramsey romances every woman in California." But privately Sarah mourned that Bill was no longer correct about Kight. He may once have been a man who played the field, but his own mother had as much as come right out and told Sarah that now he had marriage very much on his mind. As she remembered the kiss he'd given her last Monday, she felt a soft flutter in her stomach. But that kiss meant nothing. It was merely the relfex action of a man used to sipping nectar from every flower he came across. As a matter of fact, she consoled herself, he might never get over that old habit, even when Vivica had him safely married. Sarah knew that many men held their marriage vows cheaply. Kight Ramsey might very well prove to be one of them. Yes, Sarah thought with deep regret, she was certainly better off without him.

During her packing, Sarah discovered that moving time is the ideal opportunity to get rid of things. She cleaned out from her closet and bureau drawers many worn and out-dated clothes that she hadn't worn for several years but kept on the off chance that someday they'd come in handy—say, for painting a room or refinishing furniture. Now, out they went. In the bottom drawer she came across the Chinese dress Ben had given her for college graduation, which he'd purchased during a trip to San Francisco's Chinatown. Although she loved it madly, her simple life had never given her the opportunity to wear it. It was made of heavy peach silk, cut in the traditional Chinese manner, with a high neck, long narrow sleeves, and slits up the sides. The diagonally placed buttons were covered in black silk, and the only other adornment of its elegant simplicity was a bank of exquisite hand embroidery around the circumference of the hem. As Sarah tenderly repacked it in tissue paper, she thought of Ben and hoped that, wherever he was, he approved of the change in her life.

When Sarah awoke on Saturday morning she was relieved to see that the day was cool but dry—perfect weather for moving. Bill arrived at eight o'clock with a pickup truck he borrowed from a friend of his. Before he began the day's task, Sarah made him sit down to a hearty breakfast of orange juice, fresh squeezed from the tree in the apartment's common, bacon and eggs, toast, and lots of hot coffee. When he heaved himself up from the dining room table, patting his stomach with satisfaction, Sarah smiled with pleasure to see him, for once, full up to the hilt. Like most bachelors of her acquaintance, his daily diet courted scurvy.

Since Sarah's apartment had been furnished with basic furniture, she'd bought few things of her own. She and Bill made short work of what there was: several good contemporary floor lamps, her stereo components and their Danish walnut cabinet, her sewing machine and television, and a lovely antique maple rocking chair that Sarah's own mother had cuddled her in as a baby. Most of what she owned was packed in boxes, and shortly after noon, everything was on the truck.

Bill secured the tailgate of the pickup and then returned to the empty apartment to help Sarah leave it as clean as she'd found it. When all the moving debris was vacuumed from the living room carpet and all the dust mice from the corners, when the floor in the kitchen was scrubbed and the bathroom shining clean, Sarah looked around her at the eerily echoing rooms and felt a sting of tears in her eyes. Brushing them away, she wandered to the glass wall and leaned against it, gazing out at her precious little garden.

"You've been happy here, haven't you?" Bill said, coming up behind her to put his arm companionably around her shoulders.

Sarah sighed. "It was the first place that was my very own. And that garden is my last link with Ben."

"Now, Sarah," Bill scolded gently, "you'll always be linked with Ben so long as he's in your thoughts." And then with quiet understanding, he said, "Change is always hard, isn't it, even when it's a wanted change?"

Sarah nodded wordlessly and turned to give him a watery smile. She took a deep breath and exclaimed,

"All right, friend, up and at 'em! Ramseys, look out, here comes Sarah Halston!"

Laughing together, Sarah and Bill chased down a suddenly reluctant Duke, finally ambushing him in a corner of the kitchen. While Bill locked up the windows and doors, Sarah manhandled Duke into his detested cardboard cat carrier and stowed him in the middle of the truck cab's seat. She then delivered the keys to Mr. Nelson, said good-bye, and in minutes they were junketing through the city streets on their way to the freeway that led to Sarah's new life.

About a half-hour later, Bill took the exit from the swift-paced eight-laned freeway and slowed his speed considerably to safely negotiate the narrow, rough country road that led to the Ramseys' property. Sarah pointed out the hard-to-see sign and Bill swung the truck up onto the rutted driveway. When they'd reached the parking plateau, he whistled in admiration at the setting of the house.

"Quite a spread, Sarah." He let the truck idle as he looked around at what could be seen of the five acres. "You don't think you've bitten off more than you can chew, do you?" he teased.

It did look bigger than it had last Sunday, Sarah thought worriedly. "I hope not," she said, pulling a wry face. "Anyway, I'll give it my best go."

Telling Bill to wait for her, Sarah walked across the lawn to the front door, hoping Mrs. Mole was out, but no such luck. The door was opened seconds after Sarah's ring, and the housekeeper stood there with her, by now, familiar scowl.

"I'm here with my things, Mrs. Mole"—Sarah gestured toward Bill, waiting in the truck—"and I wondered if I might drive the truck over the lawn to get closer to the poolhouse?"

"Not if it's up to me, you won't," the woman said, turning her back on Sarah and disappearing into the house. Sarah stood there wondering what to do. Was the answer no, then? Or, since she'd left the front door open, had she gone to consult with Mrs. Ramsey? Sure enough, in a moment Sarah heard two voices coming nearer, and then Mrs. Ramsey was there with a welcoming smile on her pleasant face.

"You're here, my dear! Right on schedule, too. Everything is working out so well that I think it must be a good omen. I'm more sure than ever that our venture will be a success. Now, Agnes tells me you'd like to drive the truck over the lawn, and I think that's a sensible idea. Nothing could make that lawn look worse than it does now, anyway."

Sarah thanked Mrs. Ramsey, said she'd see her later, and went back to direct Bill around the edges of the lawn where the weight of the truck would harm it the least. The lawn did look terrible, but there was no point in risking tire tracks in the soggy soil, either.

The first order of business was to liberate Duke from his carrier. When that was done, Sarah and Bill spent the next several hours unloading the truck and placing boxes in the rooms where their contents belonged.

The bedroom was furnished with a double bed, and along one wall were built-in drawers and a dressing table so that the space of the room needn't be made even smaller by the use of bulky dressers or bureaus. One whole wall was of glass and looked out onto the grounds that led to the river. The drapes, a warm cream color with threads of grey and beige woven through, complemented the thick grey carpet. It was in this room that Sarah put her mother's rocking chair and one of her floor lamps. As soon as the chair was placed, Duke, with an air of relief, curled up in it for his afternoon snooze.

Because Sarah and Bill hadn't stopped for lunch, they were both ravenous by four o'clock, when the truck was empty and everything Bill could help with was done.

"I'll unpack these boxes bit by bit, Bill. Now you relax while I whip us up something for supper."

"Oh, Sarah, you don't need to do that," Bill protested. "You won't be able to find a thing in that mess. Let's run down to that little café next to the gas station we passed coming up."

"No, I've thought this all out. I packed a box with just the makings of supper in it. There's even a cooler with beer on ice."

And in seconds, Bill found himself resting on the comfortable couch in the living room, a can of cold beer in his hand. Sarah went to the cunning kitchen and looked around her, as delighted with it as a child playing house. She took from the special box the makings of Reuben sandwiches and spread them out on the yellow counters. She slathered four slices of dark rye bread with bottled Thousand Island dressing, opened a can of sauerkraut and distributed it on the bread, then topped each piece of bread with slices of Swiss cheese and canned corned beef. When the assembled sandwiches were sautéing in melted margarine on the stove, she took from the box two plates, two napkins, and flatware. Out from the cooler came a carton of potato salad she'd bought at the supermarket yesterday. In a twinkle the table was set, the salad spooned into a bowl, and the sandwiches sautéed to a buttery, crusty brown. Sarah called Bill to the table and he pronounced her a magician and set to with a hearty appetite.

After Bill had eaten his portion and, at Sarah's urging, half of hers, he leaned back in his chair and sighed. "That was a better meal than we had in Old Sacramento," he said with a contented smile.

Sarah waved away his compliment, quoting, "There's no sauce in the world like hunger."

Bill stretched and got up to wander into the living room. "This is a cozy little place, Sarah. All you need to make it perfect is a fire in the fireplace." The darkness had fallen and a winter rain was drizzling outside. Sarah agreed that a fire would suit very well. "I think I remember seeing a few logs stacked outside the house," she said.

In minutes Bill had fetched three logs and, using newspapers from the packing boxes, laid and lit a cheerful fire in the white brick fireplace. "Now I want you to make yourself a cup of coffee and sit here and relax until bedtime," Bill said, preparing to leave.

Nothing Sarah said would convince him to stay and enjoy the fire with her. "No, I've got to get up early in the morning. Pete and Jim and I are going up to Tahoe to ski," he said.

Sarah walked out to the truck with him, folding her arms around herself because of the wet chill in the air. They stood beside the open door of the cab and Sarah had a sudden flash of overwhelming affection and gratitude toward her dear friend, who had so unselfishly given her one of his precious days of rest from work. "You're always so good to me, Bill. I can't tell you how grateful I am," she said, putting her arms around his neck and giving him a friendly kiss.

Bill put his arms around her waist and hugged her. In a soft, shy voice, he said in her ear, "I'm lucky to have a friend as dear as you."

When Bill had started the truck and was headed for the driveway, Sarah turned toward the poolhouse and there in the open doorway stood a male figure, the light behind him hiding his features from her. At her involuntary squeak of fright, the man stepped forward, and with a flood of relief she saw it was Kight.

"You scared me to death!" she cried. "I wish to heaven you'd announce yourself once in a while, instead of continually creeping up on me like that!"

"Sorry," he drawled sarcastically. "I didn't want to interrupt the lovers' tender parting scene."

So he had seen her hug and kiss Bill and put his own construction on it. Well, what could be more convenient? Sarah thought meanly. Maybe it would convince him to treat her with less license if he thought she belonged to another man. Therefore, not denying what he seemed to believe, she said, "I assume you have a reason for being here, so you might as well come in so I can close the door."

"If you think you can spare me a moment of your time," he said with an ungraciousness to match her own.

Inside, Kight looked around him with an irritatingly proprietary air, noting the cozy fire, walking to the kitchen, where he saw the remains of a meal for two, and remarked, "One reason I came was to see if you needed help settling in. But I see that my offer would be superfluous. I should have realized that a woman like you would have . . . uh . . . other resources."

"Yes," Sarah said shortly. "I have." But then her natural hospitality forced her to add, "Now that you're here, may I offer you something to drink? There's the beer I had on hand for Bill, or coffee or tea. That's the extent of it, I'm afraid."

Kight arranged his long, lithe, muscular body on the couch, where the flames from the fire reflected on his brown hair, turning it a vibrantly dark coppery glow. He looked at her with a speculative smile on his firm, sensual lips. "Whatever you're having will be fine."

As Sarah returned to the kitchen to heat up the coffee, she was aware of that same uneasy but thrilling feeling in her stomach that she experienced every time she was in this man's presence. When the coffee was heated, she set two filled cups on a tangerine, laquered Japanese tray and carried it to the living room. She bent over to set it down on the coffee table in front of the couch. As she straightened, Kight suddenly stood up and grasped her wrist in his hand. "Sit here, where you'll get the benefit of the fire," he said, pulling her around the table and down onto the couch beside him. The softness of the cushions tilted her nearly into his lap, and she quickly righted herself and scooted down to the opposite end of the couch.

"Do I have something catching?" he asked, with a knowing smile.

Her answering smile was cool. "Nothing I'm not immune to, I'm sure."

Kight raised an eyebrow. "Ah, there's that peppery tongue again. You know, I really should have warned Viv that you're not a person to trifle with."

"Yes, perhaps you should have. And you'd better keep it in mind yourself."

The warm glow in his eyes faded at her words, and he replied distantly, "Yes, you may be sure I shall."

During the short time it took him to drink his coffee, they made only polite, meaningless conversation. Sarah was relieved to stop fencing with him because not only did his presence addle her brain, but she was also exhausted from the day's work and had little confidence that she had the emotional strength to resist him, if he but knew it. Her only safety lay in making sure he didn't know it!

Kight replaced his empty cup on the coffee table and stood up to go. "My other reason for coming was to relay an invitation from my mother. She'd like you to have supper with her at the house after you've finished settling in tomorrow. She said no one should have to see to her own meal after unpacking all day."

Sarah was touched and the smile she gave Kight was genuine. "How very kind your mother is. I do appreciate her invitation, but I won't be unpacking tomorrow. You remember the little girl you met at my apartment? I've promised to take her to Sutter's Fort tomorrow if the weather is good. If I don't see Mrs. Ramsey myself, perhaps you'll give her my thanks and my regrets?"

"Of course I will," he said, walking toward the door. "Do you spend all your Sundays with . . . Iris, was it?" he asked.

"Yes, I do."

"And doesn't your friend Bill object to being deprived of your company on Sundays?" Kight asked curiously.

"Of course not!" Sarah said vehemently, forgetting for a moment that Kight had the wrong picture of her friendship with Bill. "Sometimes we both take her out," she added in a softer tone. This had happened rarely, but it was true enough to satisfy Kight's obvious opinions on how a love affair ought to be conducted.

"However, I take it he won't be accompanying you tomorrow?" Kight persisted.

"No, he's going skiing with friends," Sarah replied, intending to give Kight a reason why her "lover" wouldn't be spending Sunday with her. But the dubious look on Kight's handsome face told her that for a man to go skiing with friends, instead of with his lady love, sounded even fishier than spending most of one's Sundays apart from one's lover. What a tangled web we weave! Sarah thought. Trying to seize back the control she'd lost, Sarah walked Kight toward the door, asking him again to thank his mother for her kind thought. As she held the door open for him, Sarah said, "And thank you, too, for coming by to see if I needed help."

Kight turned and gazed searchingly into Sarah's eyes, and in a tone more regretful than the occasion warranted, he replied, "I'm just sorry my help wasn't required."

So mesmerized was she by his eyes that Sarah made no move when Kight reached out and twined a tendril of her hair between his fingers, then cupped the back of her head in his palm and drew her face to his. At first, the pressure of his firm, warm lips on hers was light.

Then, as a deeply slumbering part of Sarah awoke, and her treacherous body yielded up its stiffness and swayed into his, the pressure of his kiss deepened. Sarah remained contentedly in Kight's embrace for a timeless moment, poised between the seductively beckoning golden red fire and the chill, moist night air. It wasn't until Kight's large firm hand traveled from the thick waves of Sarah's hair down to the sensitive nape of her neck that the spell broke and she surfaced from the deep well of her physical responses to this man. Her eyes flew open in consternation as she pushed against Kight's chest with all the strength in her arms and gasped, "Really, Mr. Ramsey! I thought you understood that I . . . that Bill and I . . . how quickly you forget that I'm not a person to be trifled with!"

"On the contrary, Miss Halston, that fact is uppermost in my mind," Kight replied with dry courtesy, but his lazy smile and the gleam in his brown eyes mocked Sarah's indignation. "A friendly gesture, meant only to welcome you to the Ramsey team—and how cruel you are to call it a mere trifle." And with a wicked smile and a jaunty wave, he sauntered off into the dark, wet night.

The next day was sunny and fine and Sarah was relieved that the weather wouldn't keep Iris from one of her favorite outings to the hundred-and-forty-year-old fort. In 1840, John Augustus Sutter, an adventurous Swiss, was granted 48,000 acres of this rich valley land by the Mexican government, which then owned California, in return for becoming a Mexican citizen and promising to colonize the area and preserve order. Today all that remained of that original grant was the white adobe fort standing on a knoll covering an area two city blocks long and one deep. Now, in the midst of the thriving modern city of Sacramento, it stood, an enclosed world of the past, a reminder of the romantic days of the Old West.

It was past noon when Sarah followed Iris over the broad lush lawns dotted with ancient, noble trees and through the massive, high doors of the fort. The entrance booth was just inside the door, and Iris, dressed in jeans, a red-and-white-checked shirt, and a denim windbreaker, jittered impatiently while Sarah paid the small admission charge. The two of them had been here often enough that neither any longer made use of the futuristic electronic plastic wands that visitors could carry from room to room, listening to a recorded voice relate the fascinating details of daily life at the fort.

At one end of the rectangular area was a two-story building, also of adobe brick, which housed the reception rooms and Sutter's office and living quarters. But it was the many small rooms lining the fort's inner walls, each open to the common grassy area in the middle, that fascinated Iris. Now, seeing a crowd in front of the fort's kitchen, she took Sarah's hand and pulled her across the common to join them. A young woman stood inside the small, dim interior, chatting with the crowd about the fort's food supplies as she kneaded a large batch of bread dough on a worn, splintered wooden work table. This bread, thick crusted and sturdy of texture, had been dubbed "adobe" bread back in the days when the fort was a trading post and hostelry for all manner of visitors, such as soldiers,

Indians, settlers from the east, and adventurers of all kinds. The young woman explained that when the dough was ready she would bake it in the enormous domed adobe ovens—called beehive ovens because of their shape—which stood just behind the crowd at the edge of the common. Behind the young woman was an enormous fireplace, covering nearly the entire back wall of the kitchen, with a plastic model of a side of beef skewered on the long, high spit.

Moving away from the kitchen, Iris began her visit to all the fascinating little rooms where every skill and craft necessary for survival in those rugged days had been practiced. There was a blacksmith shop, carpentry, distillery, the cooperage with its wax figure of a man molding barrel staves, a saddle shop, a weaving room, a chandler's shop where bull and ox tallow were made into candles, the chief source of light at the fort, and, of course, a gunsmith's shop. When Iris had seen all these, and the hatter's and shoemaker's shops, the grist mill, and the trade store, as well, she took Sarah's hand and pulled her in the direction of the one room that Sarah always wished, fruitlessly, that she would ignore.

Off in one corner of the rectangle was a small, dark anteroom with a doorway so low that only a child could enter without stooping. From this anteroom one went down three worn and dangerously slippery stone steps to an even smaller, darker room with a barred door. This was the fort's jail, which Sutter called the Calaboose, from the Spanish word for dungeon, *calabozo.* The room was bare except for a crude cot and an uncomfortable lifelike wax figure of a prisoner hunched in one corner of the cell, his posture one of utter dejection.

As she did on every visit, Iris stood unnaturally still, staring into the room. At these times Sarah's heart grew heavy that the concept of crime and punishment was not abstract but intimately personal to a child so young.

"They put bad people in here," Iris said, more to herself than Sarah. "Maybe he stole someone's money, or a horse. Maybe he got drunk and started a fight."

"That might've been the reason, dear." They'd been over this ground so many times that Sarah knew Iris didn't think the figure was real, even though she spoke as if she did.

Iris frowned and said, "This man probably hit his wife and that's why they put him in here. He hit his little girl, too. Do you think they still love him, Sarah, even though he was so bad and now he's in jail?" Iris asked, turning her wide, clear blue eyes on Sarah.

What could Sarah say? How could she answer for a child confronted with such an agonizing dilemma? Sarah wanted to pull the child away, but heeding Bill Blanding's instructions to let Iris express her repressed fears whenever and however she felt the need, Sarah said only, "I don't know, darling. What do you think?"

Iris sighed and turned from the door. "I think the mother and the little girl are sorry for the man because now he's all alone. But they can't love him anymore. They moved far away so he can never find them again."

The thin, bright winter sunlight seemed more intense than it was in comparison to the dim, depressing room they'd just left. As Sarah blinked, waiting for her eyes to adjust, she felt Iris pull on her jacket and whisper, "Look! There's that man who liked our cookies!"

During the precious seconds before he saw them, Sarah drank in the sight of Kight's long, graceful body leaning against the giant Turkey Oak at the end of the common. His hands, resting relaxed in the pockets of his suede jacket, contrasted with the tense alertness of his fine head as his eyes combed the area for someone—for Sarah? One had to hand it to him—he was certainly persistent, Sarah thought, annoyed, but flattered, too.

"There you are!" He covered the distance between them in several long strides. "Sacramento's finest baker," he said, squatting down to Iris' level, smiling into her abashed eyes. "I heard you'd be here today and I wanted to invite you to supper, so here I am." Then, looking up into Sarah's wary face, he said as an afterthought, "Oh, hello, Red. You're invited, too, of course."

Who did he think he was kidding? Sarah wondered. She'd thought last night, when he'd assumed a love relationship existed between Bill and herself, that he'd consider her out of bounds. But obviously it meant little to him that her affections lay elsewhere or that he was poaching on another man's prior claim. Sarah thought all this as indignantly as if her relationship with Bill were indeed what she'd let Kight believe it to be. Here he was, out of nothing but sheer habit, still pursuing her, and this time, of all the despicable methods, through an innocent child! What more did she need to prove that Kight Ramsey was nothing but a common Don Juan?

Iris looked up at Sarah with a glow of excitement in her fine blue eyes and a mute appeal that they might accept the tall man's invitation. With a mischievous smile quirking her mouth, Sarah said, "If you'd like to have supper with Mr. Ramsey, of course we'll accept with pleasure." Then Sarah turned a bland smile on Kight and added, "But first I'm sure he'd love you to show him around the fort. I imagine it's been ages since he's been here. Why don't the two of you walk around and I'll just sit here under the tree and wait for you?"

Kight, to his credit, didn't let Iris see that a tour of the fort wasn't quite what he had in mind. It was only through a small, wry smile that he acknowledged that Sarah'd turned the tables on him. Iris gave Sarah a mixed look of exhilaration and fright. Then in her funny near-adult way, she began to tell Kight her version of the recorded tour recital. Kight threw Sarah a jaunty salute and loped good-naturedly in the child's wake.

Sarah sat waiting under the immense Turkey Oak, brought over from the Swiss Sutter's German birthplace in Kandern, Baden, and planted by the Native Sons and Daughters of the Golden West in 1939, the centennial year of the founding of the fort. As Sarah lazily watched the few visitors stroll around the grounds, it occurred to her that in spite of Kight Ramsey's unpleasant characteristics—his flippancy and arrogance—she nevertheless sensed a basic trustworthiness in him that let her leave Iris alone in his company and under his influence with perfect confidence. Sarah, a woman, wouldn't trust him where she herself was concerned, but she knew he would never harm a child. And God knew Iris needed exposure to an adult male who could be trusted.

Barely fifteen minutes later Sarah saw Kight and Iris approach the entrance gate and the end of the tour. She watched as they stopped in front of the room where Sutter had received the news, in January of 1848, from James Marshall, his partner in a lumber venture, that gold had been found in the tailrace of their sawmill on the American River near Coloma, in the foothills of the Sierra, not far from Kight's country place.

Kight chose a small restaurant nearby the fort. Designed to delight any child, it specialized in pancakes with all manner of delectable toppings, light sandwiches, and extravagant ice-cream desserts. Sarah was amused to see that Iris was on her very best behavior, sitting primly on an old-fashioned soda fountain chair, her hands folded decorously in her lap.

During the meal, Kight bestowed his attention almost exclusively on Iris, which charmed Sarah, but at the same time made her feel like a fifth wheel. Finally, when they'd finished their meal, Kight casually said to Sarah, "By the way, Red, I've promised to take Iris to Coloma to see the exact spot where gold was discovered, and she informs me that she'd like you to come along."

Emboldened by Kight's gentle attentions during the afternoon, Iris spoke up in a forthright manner foreign to her where men were concerned. "Why do you call Sarah 'Red'? It's not a very pretty name."

"I think you're right, little one. I'll grant you that Sarah's a much prettier name than Red."

An uncomfortable silence fell among the three, with

Iris darting glances from one adult to the other, sensing that something had gone awry. If Sarah's own desires were all that mattered, nothing pleased her more than to hear her nickname on this man's lips. But a nickname was an intimate appellation, not to be bandied about by all and sundry. This man was not close to Sarah, nor would he ever be; she'd only be courting heartache to allow him any intimacy.

As Sarah busied herself gathering up her handbag and Iris's jacket, she said in a friendly but distant manner, "Why not just call me Sarah? After all, it is my name and I happen to like it."

When they reached the street, Iris shyly thanked Kight for the treat, and he smiled warmly at her as he bent to lightly kiss her cheek. But his manner to Sarah was cool and abrupt. In a perfunctory tone, he said, "If you like, we could take Iris home in my car and then return together to the acreage. I'd have your car picked up and brought home tomorrow."

Again Sarah felt that tearing ambivalence at his offer. How exhausting it was to be in his company! Never before had she known her emotions to be in such turmoil over a man. "Thank you for giving Iris such a lovely treat, but I really think we'd better say good-bye now."

"I see," Kight replied in a hard, flat voice. "I defer to your good judgment. Drive carefully, then, Sarah. Perhaps we'll run into each other now and again." And with a last baleful scowl, he stalked away to his car and in a blast of engine noise and a screech of tires he shot away from the curb.

"Well!" Sarah said with asperity. "If that's the way

he drives, I'm certainly glad I didn't let him drive you home!"

Iris curled her little hand into Sarah's tense fist. "I think you hurt his feelings, Sarah."

The two of them stood there on the darkened street and silently watched the powerful car as it careened around the corner and sped away, not in the direction of the highway that led to the acreage, but toward the heart of town.

Chapter Six

By early March Sarah had made a good start on returning the acreage to what it should be, although there was much still to be done. At her instigation, a crew from Ramsey Construction itself had bulldozed the driveway and resurfaced it with fresh gravel, and the front lawn had been weeded and heavily sprinkled with pre-emergent weed-killer to abort the new crop of weeds that was responding to the call of spring as exuberantly as the desired plantings were. The rose bushes were bursting forth with the fresh red leaves of new growth, an herb bed had been planted by the side door near the kitchen, and the strawberry bed on one of the terraces behind the house was heavy with still-green fruit.

Sarah had seen very little of Kight since the day of their visit to Sutter's Fort, and then only in passing. She should have been glad of this, since it was her wish to have little contact with him, but perversely, it saddened her to see him come and go with naught but a civil greeting or a perfunctory wave from the distance.

One balmy evening after a day's grubby work at clearing the weeds and dead brush from under the oleander hedge around the lawn, Sarah took a refreshing shower. As she stepped from the shower stall, she admired again the crispness of the black and white ceramic tiled bathroom, whose starkness she'd softened with touches of warm butterscotch bath linens and lacy, green ferns which thrived in the humid atmosphere of the room. As she stood in front of the mirror brushing out her thick auburn hair, she noticed that already the spring sun was turning her skin to a honey color, which, she had to admit, was not unattractive. In keeping with the languorous, almost painful tenderness of the spring evening, Sarah chose from her bedroom closet a filmy, clinging pair of lounging pajamas in pale lavender QuIana nylon. It was a shame to wear it with no audience, she smiled to herself ruefully, but it was comfortable as well as provocative.

On her way to the kitchen Sarah opened her front door, the better to enjoy the sweet aromas wafting on the evening air from the almond, plum, and apricot trees in radiant blossom a few yards from the poolhouse. Then she added to the spring-lamb stew a scant cup of newly shelled peas from her own small vegetable plot behind the house, gave it a stir, and popped it back into the oven to continue warming. A green salad of leaf lettuce, radishes, and green onions, also from her own garden, and a sourdough roll and herb butter would complete her meal.

Sarah went into her living room to put a record of Chopin études on the stereo, and over the first sparkling notes from the piano, she heard a much-missed voice from the doorway say, "I've followed my nose, and look where it's brought me."

Sarah's heart skipped a beat as she turned and saw

Kight leaning against the screen door, his head cocked and that appealing smile on his tanned face. Because she only stood there in mid-movement, like a child playing statue, and stared at him wide-eyed, he prompted her in a dry drawl: "May I come in? Or will I be interrupting your dinner?"

Springing to life, Sarah remembered her manners. "Oh, no, do come in." After all, the man was the next thing to her employer. Perhaps he had another message from his mother. But, no, it seemed not, for he came into the middle of the room with that self-assured manner of his and leisurely looked her over from head to toe, missing no detail along the way. Sarah felt a tingle suffuse her body at the boldness of his eyes and thanked her lucky stars that for once she was dressed in something other than work clothes. Not that she wanted to attract him in any way, Sarah reminded herself, but no woman wants to always look like something washed up on a beach.

"I promise I won't stay long. I'm sure Mr. Blanding wouldn't care to find me here when he arrives," Kight said with a slight questioning tone.

"Oh, he wouldn't mind," Sarah blurted. Then she remembered that she'd led Kight to believe that her relationship with Bill was such that he would indeed mind. "I mean, he won't know—he isn't coming—" But that was even worse! It sounded disloyal, deceitful! To cover her confusion and to put something—anything—between herself and Kight, she bent down to pick Duke up from between Kight's feet, where he was purring mightily as he weaved in and out at a slant, as if he were braiding himself into the man's legs.

"So you're on your own tonight," Kight said. Then

with a slightly wicked smile, he quoted from Gray, "'Full many a flower is born to blush unseen, and waste its sweetness on the desert air,' but not Sarah Halston, if I can help it." With a blandly innocent air, he added, "As it happens, I haven't had my dinner, either."

What cheek he had! Sarah thought, almost admiringly. What could a decent person say to such blatant self-invitation except: "Well, you're welcome to stay, I suppose, if you don't mind taking potluck."

With a self-satisfied smile at the success of his maneuver, he said, "From the smell of things, I'd guess that Julia Child herself would feel safe taking potluck at your house."

"Better save your compliments until you've eaten," Sarah said drily, although she knew the stew would be good. She took pride in her cooking, but it wasn't every night that she bothered for herself. Like her lounging pajamas, the stew was sheer luck. She might very well have planned to dine on canned tuna and saltines, just as she might as easily have put on her cozy but ratty old bathrobe with the torn hem that needed sewing up at the back.

But she hadn't, and there was even a portion of rich, creamy cheesecake in the refrigerator which she served Kight with after-dinner coffee. When he'd eaten the last bite, he smiled at her across the table with a measuring look in his heavily fringed eyes. "You really are a woman of many parts, aren't you? No one could ever say of *you* that you're just another pretty face."

At this comment, Sarah raised her eyes curiously to his. Was there some unspoken comparison to Vivica in his words? If so, from whose point of view? Not his, surely. Not Agnes Mole! Perhaps Mrs. Ramsey! Yes, that must be it, or, more likely, there was no comparison to Vivica at all. Nothing like wishful thinking to find hidden meaning where it didn't exist!

During dinner Sarah had sipped at a glass of white wine that had seemed never quite full or quite empty, due perhaps to Kight's generous hand in pouring it. Now she felt a relaxed expansiveness as they moved to the living room. Kight asked if he might lay a small fire to take the spring evening's chill out of the air, and while he did so, Sarah selected a record of soothing background music and put it on the stereo.

She seated herself on the sofa and watched the play of Kight's back and leg muscles through his close-fitting trousers and shirt as he bent to arrange the heavy logs on the grate. At the sensation this aroused in her, Sarah heard a little warning voice, quite like an hysterical Victorian maiden whispering in her ear, suggesting that she might be letting herself in for more than she had the will to resist.

True, Sarah *was* a maiden, and intended to remain so; but she was a modern, liberated maiden, fully capable of deciding for herself where, when, and to whom she chose to first give herself. So long as she remembered that this man's feelings for her were shallow and frivolous—and how could she forget it? —where was the danger? Surely she could enjoy a pleasant evening with an attractive man without priggishly and prudishly dwelling on assault or seduction, seeing her own downfall behind every playful innuendo or casual kiss. She'd told her share of men "no" in the past; she could tell this one "no," too. Thus spake the wine in Sarah's blood, but the inner Victorian maiden wrung her hands and pleaded, "At least get off that sofa and sit on the chair!"

Sarah rose abruptly and changed her seat to one of the armchairs facing the sofa just as Kight rose from kneeling and turned to face her with a smile.

"Now that you've just favored me with a superb dinner, I feel justified in asking you for another favor."

"Oh?" Sarah replied, suddenly alert.

"I'm going to Mexico tomorrow to bid on a dam. It's an enormous project to bring electricity to an entire region that's never had it before. The man in our company who usually does these jobs is in the hospital in traction."

"I see," Sarah murmured, not seeing at all where she fit into this impressive venture.

"Since I'll be gone as long as a week, I'd feel easier in my mind if Mother and Mrs. Mole weren't alone in the house. Would you consider moving in with them while I'm gone?"

Sarah didn't have to hesitate; she was only too glad to be useful to Mrs. Ramsey, of whom she'd grown very fond in the weeks she'd worked for her. "Of course, if it will please her—"

"Well, as to that," Kight said, smiling, "she'd already decided to ask you to take your evening meals with her, for the pleasure of your company. She likes you very much. But it might be more . . . uh, diplomatic . . . if you soft-pedaled the safety aspects of it."

Puzzled, Sarah asked, "But how else would I justify moving into her house, bag and baggage?"

"Oh, I don't mean that she doesn't *know* I'm asking

you to stay with her for safety. It's only that she thinks it isn't necessary, and it will just go down better if that aspect of it isn't dwelt upon."

Sarah nodded, smiling. "I understand. I agree that she should be looked after, and I would've done just as you did—talked her into it."

Kight laughed shortly. "It would've taken weeks to talk her into it. No, I blackmailed her into it. I just told her if I didn't have my way, I'd refuse to take the trip at all. I'd let a competitor have the job by default."

Sarah was amazed at the arrogance with which he related this typical example of bullying. "I see," she said again, faintly.

The expression on his face seemed to say that another task had been efficiently disposed of. With deliberateness he took a handkerchief from his back pocket, carefully wiped the ashes from his hands, then replaced the cloth. Four steps brought him from the hearth to stand in front of Sarah's chair. With a fluid movement too fast and smooth to allow for any resistance, Kight bent down, and placing his hands under Sarah's elbows, he drew her up as if she were thistledown.

Before she could react, he enfolded her in his arms and his mouth came down on hers in that same hawklike way she'd never forget from his first kiss. But now the tip of his tongue delicately opened her lips, and at the touch of his tongue on hers, Sarah felt a shock of hot lightning stab through her. As the length of his hard body pressed itself against hers, the heady aroma of cologne, tobacco, and something more intimate, purely male, filled her nostrils.

Moving his mouth from her lips to the sensitive lobe of her ear and down to the hollow of her throat, he murmured words she couldn't understand but responded to all the same with a weakening languor that trickled through her veins like a drug. As if from a vast distance, from a world apart, Sarah heard the sly crackling of the fire and the sighing of the orchestra's violins.

And then Kight pushed aside the silky cloth of Sarah's pajama top and with his hard, hot hand stroked the cool, smooth skin of her back and shoulders. Losing all semblance of restraint, Sarah responded by pressing with her whole consciousness closer to this man, who evoked from her feelings of physical desire that she'd never felt or dreamed of before. As his demanding but gentle hand found the softness of her breast, Sarah moaned and reached up to entangle her fingers in his soft, springy hair and pressed her mouth against his with a hunger to match his own.

Suddenly Kight's intimate touch on her body was gone and Sarah felt herself being put away from him with the same strength of purpose with which he'd embraced her just minutes ago. Bewildered, she stammered, "Kight? What's wrong—?"

Kight took several deep breaths and shook his head as if to clear it, like a surfacing diver. His voice was a low, strained mutter as he said, "Forgive me, Sarah. You're a lovely, generous woman, but there's a limit to the favors I can decently ask from you."

A blush as painful as if she'd been scalded burned Sarah's face. Was that all it had been to him—another favor? The Victorian maiden inside Sarah's mind burst into mortified tears, and cried, *See? Didn't I warn you? Now you're no different from all the other women who've granted him their favors!*

But the modern, more worldly Sarah thought scornfully: *So what else is new?* Of course the kiss meant nothing to him! Hadn't she just reminded herself minutes ago that this man's feelings for her were only those of a habitual Don Juan? Hadn't she even boasted that she could handle him? Well, then, why the ridiculous feeling that she'd been betrayed?

Making a heroic effort to salvage what shards she could of her shattered pride, Sarah smiled sardonically, and said in a throaty, sophisticated tone, "No apologies necessary, Kight. Let's not give what happened more credit than it deserves: a casual kiss brought on by all those clichés; a good dinner; a cozy fire on a balmy spring evening; sobbing violins on the stereo; and—as you so succinctly put it yourself a few weeks ago—the way of a man with a maid. Please don't give it a second thought. I certainly shan't."

At her blasé manner, the expression on Kight's troubled face went stiff and dead. His crooked smile didn't reach his hooded eyes as he drily agreed, "How right you are, Sarah. I couldn't have said it better myself."

After Kight left with a cool and distant good-bye, Sarah shed a few hot tears that evening. But by the next day, the more humiliating aspects of the encounter between herself and Kight had faded in Sarah's mind. She felt that she'd saved face by what she'd said to him, and as a matter of fact, from his point of view, much of it had been true. Sarah intended to spend no more time agonizing over what couldn't be undone. She'd simply be more careful in the future to keep her distance. For as always, it seemed, it was up to her, the woman, to define the boundaries of their relationship. Sarah could be touched that Kight had been so sweet to Iris that day at the fort, and she could be flattered that he'd asked Sarah to look after his mother instead of asking Vivica, the woman one would think he'd naturally turn to for help of that kind. As for the rest of Sarah's feelings for Kight, she resolved to put them out of her mind and get on with her own life.

Around four that afternoon a high wind came up and it began to rain, so Sarah called it a day and returned to the poolhouse. She changed into a pair of maroon slacks and a white sweater, then packed an overnight case with the few articles she'd need to spend the night in the main house. By the time she'd filled Duke's food and water bowls, the rain was pouring down and the wind was keening. Sarah threw her yellow mackinaw over her head and made a dash for the kitchen of the main house. Mrs. Mole must have been watching for her, because just as she reached the house, the door was opened for her and then slammed shut behind her.

"I hate this nasty weather," Mrs. Mole grumbled by way of greeting. "Get that wet thing off before you drench my whole kitchen."

Sarah deposited her overnight case and hung up her mack in the small combination laundry room-pantry behind the kitchen, then returned to see if she could be of any help to Mrs. Mole.

"I suppose you'd like something hot," the woman said in a tone a shade less surly than usual.

"If it's no trouble," Sarah replied.

"No, I just made fresh coffee. I may as well take the load off my feet for a bit and join you," she said, much to Sarah's surprise. She plunked down two mugs of hot coffee, then added, "Might as well shell the peas for dinner while we sit." She fetched a bowl of peas from the refrigerator, then lowered herself into a chair with a sigh.

There was something different about the woman this afternoon, Sarah thought. The scowl was still there, of course, but not so intense. Her mouth wasn't quite as pinched, her eyes less squinty, and there was an air of suppressed excitement about her, as if something wonderful were about to happen. The change in her was dramatic enough to make Sarah even more cautious than usual.

After a few silent moments while Sarah marveled at the dexterity of Mrs. Mole's fingers shelling peas, the woman suddenly spoke. "I sure hope this weather breaks by tomorrow."

Sarah responded with alacrity. If a truce was in the offing, she wanted to do her part. "Yes, so do I. There's so much still to be done in the garden."

"We'll have a little visitor tomorrow and I wouldn't like her to be cooped up in the house all day. She's cooped up enough in that apartment she lives in."

"Oh? Who's coming?" Sarah asked.

"Just the prettiest little girl in the world—that's all," Mrs. Mole replied in a playful way that sat uneasily on her sour visage.

"Your granddaughter?" Sarah guessed.

Mrs. Mole chuckled, a grisly sound, and tossed her

head youthfully. "As near as! I took care of Vivica from the day she was born, and when Bonnie came home from the hospital, I took care of her, too. I'm as close to the Harrington family as non-blood can get," she said proudly.

"Harrington?" Sarah echoed faintly. "Vivica has a little girl?"

"That she does," Mrs. Mole said smugly. "And the very picture of her mother she is, too." Leaning forward confidentially, she added, "It's not my way to speak ill of the dead, but all the same, it's a blessing the child took nothing from her father in the way of looks."

"Oh?" Sarah said, inadequately.

Mrs. Mole nodded vehemently. "He was one of these foreigners. Swarthy. Short. I once saw a photograph of him when he was a young man, and even then he wasn't much to look at, if you ask me."

"And he's dead, you say?" Sarah asked.

"Died about eight years ago, before the baby was even born. And died *happy,*" Mrs. Mole said pointedly. "Most girls wouldn't give a sick old man like him a second look, much less marry him and nurse him through his last illness. That girl's a saint." Mrs. Mole sighed, shaking her head in wonder.

"I see," Sarah said thoughtfully. "I hope he left Vivica well provided for." She felt a slight shame for her unworthy thought—but only slight.

Mrs. Mole made a casual, dismissive gesture. "More money than you could shake a stick at. And only proper, considering the devotion that girl showered on him. She wouldn't like to hear me say it, she's such an angel, but *I* say that a good sum of money and that beautiful child were little enough in return for everything she did for him." After a pause, Mrs. Mole added with satisfaction, "To look at the child, you'd never guess she had a foreigner for a father."

"Well," Sarah breathed, her mind teeming with this entirely new picture of Vivica Harrington.

Mrs. Mole murmured fondly, "Dear little Bonnie. I'm to have her for as long as a week, while Vivica's in Mexico."

Thunderstruck, Sarah repeated, "Mexico? Vivica went to Mexico?"

The older woman favored Sarah with a slyly triumphant smile. "That's right, with Mr. Kight. I shouldn't be surprised if they don't come back married." Then she lapsed into a watchful silence, as if to let this latest bombshell sink into Sarah's mind.

Which it did, like a ton of lead. That Vivica was a widow with a child about Iris' age was enough to digest in one setting, but that she'd gone to Mexico with Kight! As all the nasty implications of the news swarmed into Sarah's mind, she felt a sick headache coming on, and, worse, the threat of tears. Sarah's self-scorn was immense as she remembered how she'd preened like a flattered fool because Kight had chosen her instead of Vivica to look after his mother. And it had never even crossed her tiny mind that the reason for it was that Vivica was going to Mexico with him. Heavens! What a contemptible idiot she was! And that lovemaking—that had happened only after he'd gotten her to agree to help him out, just as the first kiss had happened after he'd got his mother to coax Sarah to take the job.

It all fit the pattern he'd set the first day she'd met him: to let her think whatever idiocy came into her head, so long as it suited his purposes; to trample roughshod over her feelings so long as he got his own way. Now she no longer even believed that his feelings for Iris were genuine, but only an even more evil way to manipulate Sarah through her feelings—and for using Iris, Sarah would never forgive him.

Sarah stood up. "I wouldn't be surprised, either, if they're married when they return. In fact, I hope they are."

Mrs. Mole looked at Sarah suspiciously. "You do?"

"Yes, I do. I think they're very well suited to each other. In fact, one might say they deserve each other."

"Well, now there you're absolutely right," Mrs. Mole agreed enthusiastically. "They've been close since they were children, what with their fathers being partners, and they should've married long since. They would've, too, if it hadn't been for that foreigner . . ." Her voice trailed off. Then, staring into space, she said in a wistful tone, "We'll be a family again when they marry, and things will be as they used to be . . . no more of this camping out in other folks' kitchens . . . never knowing which end is up, what's around the next corner. . . ."

Sarah broke gently into Mrs. Mole's daydreaming to ask, "Will you show me which room I'm to use, Mrs. Mole? I'd like to rest a bit before dinner, if you don't mind."

There was a bottle of aspirin tablets in the bathroom near the spare bedroom Sarah was to use. They must have helped enough to let her sink into a restless, uneasy sleep, for it was nearly seven o'clock when she awoke to the thrashing of the rain against the window and the howling of the wind. Listlessly, Sarah sponged off her face and brushed her hair before she left the bedroom to have dinner with Kight's mother, as she'd promised she would.

Mrs. Ramsey was sitting in a deep lounge chair by the fire, dressed in a pale blue brocade dressing gown that set off her soft white hair. She drew in her breath sharply when Sarah walked into the room.

"My dear child! Are you ill? You look awful!"

Sarah forced a smile to her lips. "It's only a head cold."

"After dinner you're to go straight to bed. Or would you rather go to bed now and have your dinner on a tray?"

When Sarah insisted that she'd prefer to dine at the table, Mrs. Ramsey subsided, but all during the meal she kept an eagle eye on Sarah as she chatted of this and that. When she'd been brought up to date on what had been accomplished on the grounds and what was scheduled next, a little silence fell that Sarah, in her misery, felt too sick and dull to bridge.

When Mrs. Mole brought them their tarragon chicken and returned to the kitchen, Mrs. Ramsey asked, "Did Agnes tell you Vivica's little girl is coming to stay with us for a few days?"

"Yes, while her mother's in Mexico," she said.

Mrs. Ramsey sighed. "It was all very sudden. Vivica's always been an impulsive girl, I'm afraid. This trip was planned on such short notice that there wasn't time to arrange for a proper baby-sitter."

Sarah said noncommittally, "Mrs. Mole seems delighted with the arrangement."

"Oh, yes, she dotes on Bonnie." Mrs. Ramsey paused briefly. Then, as if she'd made up her mind to say something, she continued: "Sarah, I want you to feel free to come to me if the child gets in your way or interferes with your work. She's a nice enough child, but she's never been encouraged to develop her own inner resources." She looked at Sarah meaningfully.

Sarah smiled and nodded. "Children aren't strangers to me. I know they can be pests now and then, but I'm sure she'll be no trouble to me."

With a wry grimace, Mrs. Ramsey laughed. "Don't be too sure!" Then she sighed again. "At least it will be a pleasant change for poor Agnes. She's not really happy with us here, you know. She's used to being at the center of a busy, social family. We're much too dull here for her taste."

"I wonder why she doesn't find a more congenial position, then," Sarah said, for the sake of saying something, even though she knew all too well the answer to her own question.

Mrs. Ramsey hesitated. Then in a quietly kind tone, she explained: "Well, you see, Agnes came to the Harringtons as a very young girl, just out of school. She'd been with them about five years when she married. Just months later her husband was killed in an industrial accident, so there were never any children. All she's ever had in the way of a family was the Harringtons. Now, except for Vivica and Bonnie, that's all gone, too. George, Vivica's father, died several years ago, and just recently Vivica's mother remarried and moved to Palm Springs. Vivica lives in an apartment with housekeeping service, and she feels that she simply doesn't have the room for Agnes, and even if she did, she doesn't need full-time help." Mrs. Ramsey paused, as if to examine her next words; then she continued. "Sarah, it can't have escaped you that Agnes isn't . . . well, an easy person to get along with. Finding a new position wouldn't have been easy for her, so, when the family scattered, there just wasn't anywhere for Agnes to go."

The troubled look on Mrs. Ramsey's kind face made Sarah wish she'd never asked the heartless question that had prompted these revelations about Mrs. Mole's private sorrows.

When Mrs. Ramsey spoke again, her voice was determinedly bright. "Then Kight bought this place and Agnes offered to help us out until . . . until either Vivica marries again and has room for her, or until Kight and I get ourselves sorted out and find another housekeeper." She smiled and, with the utmost tact, said, "I think we've all benefited from what could've been an unfortunate situation."

Sarah would have bet her life that Mrs. Ramsey would never lift a finger to replace Mrs. Mole until she herself expressed a desire to leave. It was quite clear now why Mrs. Mole saw Sarah as a threat to her future security. Who could blame her for trying to drive Sarah away since that very first day? She couldn't have known that Kight himself would accomplish the task for her.

Well, it was all water under the bridge now, Sarah thought wearily. She was glad she knew the story, as sad as it was, because now that she understood Mrs. Mole's motives, Sarah could assuage her fears. She'd need to be mean-minded indeed to engage in a cold war with a woman as alone as Mrs. Mole. Sarah, too, knew what it was to be alone and rootless in the world—now, more than ever, because the dream she'd barely dared to hope for had come briefly to life, only to be brutally slain.

Chapter Seven

The next few days revealed a change in Mrs. Mole's demeanor that was nothing short of miraculous. During little Bonita Harrington's visit, gone was the depressed shuffle of Mrs. Mole's footsteps about the house, replaced by a cheerful, spring gait. When her face wasn't wreathed in actual smiles, it at least bore an expression of calm contentment. Even more remarkable, the pleasure she took in the child's presence seemed to permeate her spirit so entirely that, uncontainable, the overflow extended to everyone in the house, even Sarah. What's more, it was a testimonial to the power of love that this particular child could cause such a remarkable about-face in a person of such rigid character as Agnes Mole. For Bonita Harrington, in Sarah's opinion, was not an easy child to love.

Bonnie had the looks of a Botticelli angel and the personality of an apprentice demon. It seemed true enough that she was entirely her mother's child, for she had not only her mother's porcelain-doll beauty, but also Vivica's imperiousness, made even more intimidating by the child's tendency to launch into wild tantrums at the slightest provocation. During the week

Sarah witnessed many such scenes, all of which were handled by Agnes Mole with a loving forbearance that would have been admirable even in a hardened kindergarten teacher, not to speak of a somber-natured childless woman in her fifties.

But one such scene in particular made a sharp impression on Sarah, because the two creatures of whom she was most fond in all the world, Iris and Duke, were involved. It was on Saturday evening, after Bonnie had made a shambles of dinner, that Mrs. Ramsey said gloomily, "Kight mentioned that you have a little friend just Bonnie's age."

"Yes," Sarah said, hoping she was wrong about what she guessed was coming next.

"I told you Bonnie would be bored here. Maybe if she had someone her own age to play with . . . what do you think?"

And so, reluctantly and against her better judgment, Sarah agreed that tomorrow—the day both she and Mrs. Ramsey fervently hoped would be the last day of Bonnie's sojourn—Iris would spend the day at the acreage.

The next day the little girls hadn't been together an hour before Sarah's worst fears were realized—the children's respective natures had the proverbial antipathy of oil and water. Feeling guiltily responsible for subjecting Iris to Bonnie's tyranny, Sarah resolved to stay within earshot of them all day. After a hectic picnic lunch on the lawn, Sarah left the little girls to eat their dessert of pink lemonade and coconut cake while she returned to the task of pruning the photinia hedge on the west boundary of the lawn.

The first noise Sarah heard was a strangled, guttural gasp, followed by a piercing shriek from Iris. "Stop that! Put him down!"

Then Bonnie's imperious voice said, "Shut up and mind your own business! I'm the mother, and he's got to drink his juice."

Sarah turned to see Duke's head caught tightly in the crook of Bonnie's elbow, his orange body hanging beneath her arm, his hind legs flailing desperately against thin air. Iris stood with her hands clasped rigidly in front of her chest, while her feet did a frantic, agitated dance on the meadowy lawn. Bonnie, with perfect aplomb, ignored all this travail while she poured lemonade down the throat of the sputtering, choking cat.

With an edge to her voice that cut through to both little girls, Sarah called out, "That will do, Bonnie!" With a defiant grimace, Bonnie grabbed Duke by the skin of his back and flung him as far as her limited strength allowed, whereupon he streaked away like the orange tail of a comet.

She shouted as Sarah, "I don't want to play with that stupid, ugly cat, anyway!"

Perhaps from relief and recovered security, Iris burst into tears and wailed, "He is not stupid and ugly!"

"Yes, he is, and so are you!" Bonnie turned on Iris with the mad fury Sarah had grown to dread this past week. "In fact, that's why your mother *really* left you!" she jeered. "You're so stupid you think she's in the hospital because she's sick! But *I* know she ran away and left you because you're so ugly! You're so ugly not even your own *mother* loves you!"

The look of shocked revelation on Iris' face drove

everything out of Sarah's mind but the determination to remove her from yet another destructive situation. She reached out for Iris, intending to put space and locked doors between her and Bonnie, and it was like a mallet blow to the heart when Iris flinched away from her and hid her weeping eyes in her hands. Oh, my God! Did her beloved little friend think that Sarah would ever lift a hand to her in anger?

"Iris, sweetheart . . ." Sarah beseeched.

Suddenly a deep male voice called out, "I'll take care of that one. You take Iris to the poolhouse."

Sarah looked up and there, striding across the lawn from the direction of the parking area, with packages under one arm and pointing at Bonita with the other, came Kight Ramsey with a stern, dark look on his face. Sarah picked Iris up in her arms, and the little girl, now recovered from the reflex action that had made her flinch from any raised hand, wound her thin arms tightly around Sarah's neck. The last thing Sarah heard as she headed for the poolhouse was Kight's grim voice saying, "All right, young lady, not another word out of you. Now get into that house. March!"

A few minutes later Kight appeared at the door of the poolhouse, still laden with packages, to see Sarah on the living room sofa with a small crumpled heap in her lap. Catching Sarah's eye, he let himself in. "I can't get her to stop crying," Sarah said miserably, her own eyes wet with tears.

Kight sat down beside the two of them where he could see Iris' face. Speaking to Sarah in a gravely serious tone, he said, "I suppose you didn't know that Bonnie has something terribly wrong with her eyes?"

Bewildered, Sarah gaped at him. "She does?"

He nodded sadly, taking no notice of Iris, who had momentarily stopped sniffling, the better to hear the grown-ups talk. "We seldom speak of it," Kight went on, "but I see now that someone should've mentioned it to you."

"My goodness," Sarah said, completely at sea as to where this was leading or what it had to do with the present situation.

"It's a rare disease, although not unknown in the annals of medicine, of course."

"I see," Sarah said dubiously.

"The thing is, the child is unable to see beauty."

"Ah . . ." Sarah breathed. Iris sat so still she might not even have been in the room.

Kight shook his head hopelessly. "It's very sad. She can be two inches from the most beautiful flower, but all she sees is a weed. Or give her a pretty new dress, but all she sees is a hand-me-down. Imagine going through life like that."

Playing it perfectly straight, Sarah replied, "Why, that's tragic. Poor little Bonnie."

Kight sighed. "Yes, it's an awful burden for a little girl to bear, not to ever see the beauty all the rest of us see. Of course, we never speak of it to Bonnie. What would be the good of pointing out things she's unable to see?"

"Yes, I certainly agree," Sarah said. "It's not only pointless, but cruel, too."

"Exactly. That's why when she calls something ugly, well . . . we just ignore it. After all, she simply can't *see* beauty. There's no point *arguing* with someone like that."

Kight and Sarah fell silent. Iris suddenly shuddered

from spent sobs, sat up straight on Sarah's lap, and wiped her runny nose with the hem of her knit T-shirt.

"She shouldn't have been mean to Duke," Iris said in the manner of someone willing to forgive, perhaps, but not forget.

"Absolutely not!" Kight said, acting as if he'd just noticed Iris' presence. "It was very naughty of her; no one denies that. And she'll be punished for it, too."

"Humph." With a stuffy, placated air, Iris climbed down from Sarah's lap and took a proper seat in the chair across from the sofa, her reddened eyes straying to the packages around Kight's feet.

"Now, about these packages," Kight began, launching into some rigamarole about his shopping expedition in a small Mexican town that Sarah'd never heard of and whose name she couldn't have reproduced even at gunpoint.

As she watched Kight talk so animatedly to Iris, Sarah felt a sting of shame remembering her cynical thought that Kight courted Iris only to get at herself. Sarah was fervently thankful that she'd never put the ugly thought into words. No one who'd witnessed the graceful and tactful kindness he'd shown to Iris today would ever doubt that his feeling for her, and no doubt for all children, was genuine.

A deeply warm and secure trust of this man settled into Sarah's heart. No matter what the past held, or the future, she would always remember that he'd acted the White Knight to a vulnerable, hurt child; that he'd so swiftly staunched what might have been, for Iris' fragile ego, a crippling wound.

Kight handed Iris a medium-sized package and indicated that she should open it. All the while she did so, with slow, careful motions so as not to tear the thin, gaudy paper, he related the details of his purchase.

"I went into dozens of shops and looked at hundreds of things, some of them nice enough, I suppose, but not quite what I had in mind. I was just about to give up when I came across a little shop called La Belleza, The Beauty. I thought, well, if *this* isn't the place, then I'm just sunk. I said to the woman in the shop, 'I'm looking for something worthy of a young lady who has skin as fine as a white rose petal, hair like maple sugar, and eyes the blue of a California summer sky—in short, a very beautiful young lady.' The woman put her finger to the tip of her nose and thought a minute. Then she winked at me and went to the back of the shop, and that is what she brought out to show me."

Kight finished his recital just as Iris drew from the tissue paper what Sarah recognized as a very expensive child-size version of the eyelet lace gown called a Mexican wedding dress. Iris clutched the heavy cream cotton dress to her chest and gave Kight a trembling, watery smile.

"But . . .? Is this dress for Bonnie?" she asked uncertainly.

"No! It's for you, little goose! What have I just been telling you?" Kight protested, laughing.

Iris lowered her eyes and mumbled, "I'm not *ugly*, I know that. But I'm not . . . beautiful." But there was the slightest hint of a question in her voice.

Kight smacked his forehead with the palm of his hand. "Call the doctor, Sarah! We've got to have this child's eyes examined! I think she's caught Bonnie's affliction."

Sarah saw that Iris was hovering between laughter

and tears, so she got up, gave the child a hug, and pushed her gently toward the bedroom. "Of course it's for you, darling. Now run along and try it on for us."

When Iris left the room, Sarah turned to Kight with tears in her eyes. "This has been a day that Iris will never forget. I want to thank you for making it a cherished memory instead of material for nightmares."

"Don't give me too much credit," Kight replied quietly, "because I haven't spoken an untrue word; she *is* a beautiful child, if only one has the eyes to see it."

Sarah felt an overwhelming urge to embrace Kight and hold him close forever. It was a very different urge from the one she'd felt just a week ago, right here in this very room. In fact, it was deeper, and even stronger—and just as unwise to act on, she reminded herself.

As Sarah and Kight gazed silently at each other, their faces floated closer and closer until there were mere inches between them, then only warm, sweet breath, then nothing at all to keep their lips from meeting in a kiss.

After lingering a moment, Kight's mouth left Sarah's to brush lightly against her heated cheeks, then descended to kiss the hollow of her throat. With a delicious shiver Sarah felt his arms surround her and hold her tenderly as his head lay sweetly heavy against her breast.

"Your skin smells like peaches in the sun," he murmured. Sarah felt her body slacken as that strange feeling of abandon stole through her blood, weakening her will.

At just that moment Iris bounced back into the

room, her plain little face gleaming with prideful joy. Sarah instantly and guiltily pulled away from Kight, feeling like every kind of fool to have once again lost her wits at the crook of this whimsical man's finger; but Iris seemed to notice no irregularity as she posed shyly in front of the two adults and waited breathless for their comments. And she did look like a pure little mountain flower in the white dress. When a suitable fuss had been made, Kight gave her the second package.

"This is to commemorate the day we met," Kight said as Iris drew from the wrappings a wildly colored wooden puppet, dressed like a clown. "I just wish it had a purple mouth," he said, sharing a private smile with Sarah.

As Iris manipulated the strings and laughed delightedly at the jerky motions the puppet made, Kight handed Sarah the last package. "And this is for you."

When the gift lay opened on her lap, Sarah was rendered speechless. How to respond to a gift that, under the circumstances, might have great significance—or, on the other hand, might have none at all. Paralyzed by confusion, Sarah could say nothing for the moment.

Iris had no such dilemma. Squealing with pleasure, she cried, "It's just like mine, Sarah! Now we can be twins!"

Sarah smoothed her hand lovingly over the demure smocked bodice, the graceful belled sleeves, the intricate handwork of a gown just like Iris's, a Mexican wedding gown.

In the days following that eventful Sunday, Sarah thought many times about the lovely dress that now hung in her closet waiting to be worn on the proper occasion—whatever that might be. Sarah felt like a prize fool to even conjecture about the impulse that made Kight bring her a gift at all, let alone that particular dress. Nothing in his behavior past or present would cause a sensible person to find hidden meaning in the gesture. He was a whimsical man who, in spite of his fanciful recital to Iris, had, on a generous impulse, bought a gift for a sweet little girl, and while he was at it, bought the same gift in a larger size for the young woman who, in his mind, was connected to the child. Without a doubt, it was as simple and meaningless as that. And yet, some perversity in Sarah's mind kept worrying about the incident until she feared she'd drive herself into a neurotic state over it.

Therefore, it was a relief, of sorts, when she did discover the true source of the Mexican wedding dress. It happened on an unseasonably warm day in early April. The temperature was in the eighties and Sarah found it hot, sweaty work, clearing off the masses of dead growth on the gentle slope that lay between the pool and the tennis courts. The spring air was peacefully quiet but for the somnolent sound of the bees going about their daily chores. The monotony of her work and the placidity of her surroundings had put Sarah into such a mindless state of reverie that she nearly jumped when she heard her name called out.

"Miss Halston! Hello, there! Working away like a beaver, I see."

Vivica Harrington stood midway on the terrace steps between the pool and the courts, dressed in a metallic-blue bikini so skimpy that every plane and angle of her body were there in plain sight for anyone to see. Sarah, naturally, was dressed in her working denims, a cotton shirt, and had a sweatband wound around her forehead.

"I hope it won't put you off if I sunbathe while you slog along," Vivica said, indicating the lounge chairs on the deck around the pool.

"Of course not. Why should it?" Sarah replied evenly. "I see you already have a good start on a tan."

"You can't believe how divine the weather was in Mexico. Nothing but sun and moonlight for a week straight."

Sarah knew she was meant to imagine the two of them, Vivica and Kight, together in that moonlight, and of course she did. Vivica had a glass of something iced in her well-manicured hand which she sipped at when she'd settled herself into a chaise, her face offered to the early afternoon sun.

"I came out today to have a visit with Grace, but she's up to her ears in plans for her party. Agnes is all wrapped up with Bonnie, so I'm on my own," Vivica said, sounding aggrieved.

"Mrs. Ramsey will be well organized by June if she's starting now," Sarah said, forced to respond out of courtesy.

"Oh, not that party," Vivica replied idly. "She's having a housewarming party in a week or two. Surely she told you?"

"No, but there's no reason why she should have," Sarah said, heaving a particularly deep-rooted weed from the rich soil.

"So *I* would've thought, but she *insists* you're to be invited." There was no mistaking the petulance in

Vivica's tone. "I'm sure you'll get your summons soon."

Sarah made no reply, but thoughts aplenty were racing through her head. Why should Mrs. Ramsey want hired help to come to a private party? Should Sarah refuse? Would Kight be there? If Sarah accepted, what would she wear?

As if Vivica had read her mind, she said, "It will give you a chance to wear your new dress. It's just the thing for a Sunday open house."

"You know about my new dress?" Sarah asked, surprised.

Vivica laughed, not very pleasantly. "I should say so! I nearly walked my feet off for that dress, and for that homely little girl's gifts, as well."

Sarah caught her breath. Surely Kight hadn't called Iris homely? No, Sarah would never believe it. It must have been Agnes Mole. Agnes had been kind and affectionate to Iris that eventful Sunday, but in contrast with her beloved Bonnie's beauty, Agnes wasn't likely to have seen Iris's looks as anything but ordinary at best.

Suddenly feeling totally exhausted, Sarah sank down upon the earth and took a grimy handkerchief from her back pocket with which to wipe her moist face. So Vivica had picked out the dress Sarah had spent so many foolish moments dreaming over. It had probably even been Vivica's idea. She must have been shopping for gifts for Agnes Mole and Bonnie, and while she was at it, said, "Why not pick up something for that what's-her-name, that girl gardener who's been so obliging about your mother?"

Eerily, Vivica again echoed Sarah's thoughts.

"Men," she laughed with soft indulgence, "they never know what to buy or where to find it. And it was such a hot day I thought I'd melt!"

Without much enthusiasm, Sarah voiced as much of a thank you as she could muster. "It's a lovely dress."

Into the silence that befell the two women, Vivica suddenly asked, "What do you plan to plant there where you're clearing out?"

Surprised that Vivica should show any interest in the grounds, Sarah replied, "Why, xylosma, I thought."

"I think I'd prefer those bushes with the red berries."

She'd prefer? But aloud, Sarah said, "Red berries. Now let me think. There are so many bushes with red berries. Cotoneaster, perhaps? Celastrus? Pyracantha?"

"That's it—the last one. I think I'd like that; it's so cheery in the fall."

Sarah hesitated, suddenly not sure of her position. "I think the bees would be a problem with pyracantha. They're mad for it in the spring when it blooms. And with people around the pool, I think something else . . ."

Vivica sat upright and shaded her eyes with her hand. Looking fixedly at Sarah, she said, "Miss Halston, when you own your own little plot of ground, you can plant what you like. In the meantime, please remember you merely work here, will you do that?"

Sarah's face froze and her body went rigid with insult and anger as she rose to her feet and prepared to depart. She remembered Mrs. Mole's fond hope that Vivica and Kight would return from Mexico a married couple. That hadn't happened, she knew, or Vivica would have lost no time announcing it. But obviously the marriage was imminent, for Vivica spoke as if she had the rights of a wife and property owner. Well, then, let her have the thorny pyracantha bushes by her swimming pool. By next spring when the bees came in droves to the creamy blossoms, Sarah would be long gone from this place.

"Oh, you're leaving?" Vivica said in false surprise. "I do hope it wasn't anything I said." As Sarah stalked off, Vivica sank back comfortably in her chair and once again turned her face to the sun.

Chapter Eight

During the next week things went from bad to worse for Sarah. Poor Duke, his psyche barely recovered from Bonnie's forced feeding, developed a stomach complaint and had to be taken to the vet to be dosed for hair ball. Cut worms had demolished Sarah's entire planting of green beans, and the ivy bed was infested with snails. Bill called one evening to say that the housemother for the Friends of Childhood shelter house had resigned because of illness and a replacement had to be found fast. Because Sarah had been so pleasantly surprised at Mrs. Mole's knack with children, she thought of offering her name to Bill, then immediately discarded the idea. With Vivica's union with Kight so imminent, Mrs. Mole's devoutly desired future was all but accomplished.

But, worst of all, the Reillys were losing their home. On Thursday afternoon, Sarah traveled into the city to take Iris shopping for summer clothes. Even though she was welcomed in as usual and offered a glass of iced tea, Sarah immediately sensed the air of worry and defeat in the small house. While the three adults waited for Iris to return from school, Mike told Sarah what had happened.

A commercial developer had bought up most of the neighborhood, and the Reillys' landlord had given them ninety days' notice to vacate. "Some of our neighbors have already found new places, but we've seen nothing we can afford that's big enough for the three of us in a neighborhood near a school," Mike said dispiritedly.

Maggie murmured mournfully, "I'm afraid we'll lose the child."

"Now, stow that, woman!" Mike barked at her, and Sarah realized it disturbed him to hear his wife voice his own fears. "No matter what, *that's* not going to happen."

"Of course it isn't," Sarah soothed. "I'll help you look for a suitable place. There's bound to be *somewhere* in this big city—or maybe Friends will increase your stipend—or I can supplement it myself. Or . . . I know!" she said excitedly. "The shelter needs a new housemother! Why don't you look into that? The two of you would be perfect—"

Mike interrupted gruffly. "We've already offered and been turned down. They said we were too old."

Maggie demurred. "They didn't quite say that, dear."

"As good as!" Mike snapped back. "However they put it, the answer was no." He subsided into a dark funk that wrung Sarah's heart. Mike was undoubtedly right; the organization had strict employment regulations, and not hiring anyone over sixty-five was bound to be one of them. Such a shame to waste people with such valuable experience, but there it was.

Sarah said with a briskness she was far from feeling, "Well, it's a bother, but nothing we can't remedy if we all work together. You'll see, in a week or two we'll have it all sorted out."

But later, when she'd collected Iris and was driving the little yellow car toward Sacramento's central shopping district, Sarah didn't feel so sanguine. The Reillys could easily find a home in one of the several low-cost housing developments for senior citizens, but children weren't allowed there. Or they could find an affordable apartment in certain areas of the city, but these were usually in neighborhoods far from a school. The combination of elderly people on a small income and a young child in grade school was one that was too unusual to fit into the cultural patterns of a large city. Finding a solution would require all Sarah's ingenuity.

When the shopping expedition was finished, it was nearly four o'clock. An exhausted Sarah and a beamish Iris rode from one end of Sacramento's lovely tree-lined shopping mall to the other in one of the open-sided, colorful jitneys that traversed the mall for the convenience of weary pedestrians.

When they'd retrieved the car from a carpark near the last end of the mall and were driving home on a one-way street, Sarah noticed a Ramsey Construction site on the block ahead. It was an unsettling sensation to see that name on a huge billboard looming near the edge of the site. She slowed down to point it out to Iris.

"Please, can't we stop?" Iris pleaded. "Maybe we'll see Kight."

"No, sweetie, he won't be here. He has men who watch over things for him. But we can stop, if you like," Sarah said, because she was so sure the owner of an international company wouldn't be on the site himself. Besides, it would be educational for Iris to see the inner structure of what would soon be a finished building. Around the next corner she found a place to park and she and Iris got out to stand outside the metal fence that protected bypassers and sidewalk superintendents from the dangers inherent at any heavy-construction site.

"You see"—she pointed out to the interested child—"all the workers wear what are called hard hats to protect their heads from falling objects."

Iris stood gaping about, then suddenly pulled on Sarah's skirt. "Look, there's some women!"

And so there was, much to Sarah's amazement, several women workers on the site, dressed the same as their male counterparts in heavy-duty work clothes, the required hard hats hiding whatever distinctions a feminine hairdo might have given them.

"Will wonders never cease?" Sarah breathed. She wondered how long these women had held their jobs. Had she possibly misjudged Kight Ramsey's attitudes toward women in the working world? Or had her own occupation played any small part in changing his mind? She'd have given a lot to know the answer to that question.

Over the racket of a nearby jackhammer and the rumble of a bulldozer, Iris suddenly shrieked out in ear-splitting volume, "Hey lady! Lady, look down here!"

A sturdy young woman in the cab of the bulldozer looked down from her powerful throne and, grinning widely, threw Iris a hearty salute. Iris turned to Sarah with shining eyes. "When I grow up I'm going to work for Kight, too."

"It certainly would be grand to help construct a building," Sarah said enthusiastically. They had wandered perhaps a half-block from the car, and as they began to walk back, Sarah heard a low wolf whistle and looked up to see a young worker leaning insolently against the fence a few yards in front of them.

"Hey, baby," he called suggestively, "you got plans for tonight?"

Lowering her eyes, Sarah took Iris's hand and pulled her to the opposite side of herself from the man, then hurried her steps toward the car.

"Dump the kid, baby, and you and me can have a party." To Sarah's horror, he began to move toward a gate that led to the sidewalk where she'd have to pass on her way to the car. In an urgent undertone, she said to Iris, "Pay him no attention. He's just a young smart aleck. If we ignore him, he'll get bored and go back to work."

"Sarah," Iris questioned in a quavering, rising wail, "is he going to hurt us?"

"Of course he isn't," Sarah replied firmly.

Then the man said something unbelievably coarse and ugly, and Sarah abandoned all pretense of dignity, breaking into a run and dragging a whimpering Iris behind her.

They'd nearly reached the car when Sarah heard a terribly snarling outburst of profanity. She whirled to see the young man turn tail and run, with Kight Ramsey, a look of cold outrage on his face, bearing down on him. In several long strides, Kight grabbed the man by the collar of his shirt and spun him around. As if the violent moment were being filmed in slow motion, Sarah saw every inch of the swing of Kight's heavy fist as it moved up from the vicinity of his hip and connected cleanly with the man's right jawbone. With dreadful clarity Sarah heard the man's teeth crash together and the terrible thud of bone on bone.

The man crumpled to the dirt as workers ran from the near distance to gather in a semicircle around him. In the silence that fell, Sarah heard a small, low moan nearby and turned to see Iris, her poor little face the color of wallpaper paste, collapse into a heap on the hard sidewalk.

Sarah's muffled scream brought Kight to her side in seconds. He scooped up the limp child in his arms and with a curt jerk of his head indicated that Sarah was to follow him back through the wire mesh gate. As they headed toward the temporary office shed in the middle of the site, they passed by the group of workers. The young man was shaking his head and trying to gain his feet. In a harsh, cold voice, Kight said to him, "Pick up your pay, get the hell out of here, and never bring your foul mouth onto one of my sites again. Is that clear?" Avoiding Kight's eyes, the young man nodded sullenly.

After a short exchange with his foreman, Kight bundled Iris and Sarah into the white Lincoln Continental, greyish now from construction dust, and drove purposefully through the city streets. Within fifteen minutes Sarah found herself twenty floors up in Kight's penthouse condominium. Kight ushered Sarah in and carried Iris to a large white sofa and settled her into its soft downy pillows.

"Now you lie still for a bit," he said to the conscious

but groggy child, who responded with a small, obedient nod and a wan smile. He turned to Sarah and said irritably, "I've never known a child as disaster prone as this one."

Sarah sighed. "You don't even know the half of it."

"So I surmised," he replied shortly. "But now I intend to know the whole of it. Make yourself comfortable and I'll be right with you."

When he'd stalked out of the room for Sarah knew not where, she walked to where Iris lay, saw that her eyes were closed, then wandered vaguely over to the window that was the whole south wall of the apartment. The building looked out over Capitol Mall, the long, wide boulevard that led to the golden-domed State Capitol building. Off to the right was the pumpkin-colored bridge spanning the Sacramento River, and ahead lay a view of the entire south area of the sprawling city.

Turning her back on what, at another time, would have been an exhilarating sight, Sarah headed for a soft club chair, sank down on it, lay her head back, and closed her eyes. She admitted to herself that, given her choice of anyone in the world to be with at this moment, she would choose Kight Ramsey. In her mind's eye she saw the way he'd twice appeared out of nowhere, like an avenging angel, to save Iris, and now, herself. She wanted nothing more than to turn all her worries and woes over to him for solving. But that was an impossible desire that would never be realized. Not only would he soon be married to another woman, but the sting of finding out the true source of the Mexican wedding dress had forcefully reminded Sarah that her original assessment of Kight's feeling for her was correct: that of an employer mindful of mollifying a touchy employee—with a little of his habitual "the-way-of-a-man-with-a-maid" syndrome thrown in for good measure. And Sarah had been a fool, as usual where he was concerned, to think differently even for a moment.

As Kight reentered the room bearing a tray of drinks, he announced, "I've arranged for your car to be delivered to the acreage. I'll drive you home later."

Perversely, the very quality in Kight for which Sarah had just privately acknowledged admiration—his self-confident resoluteness—now got her back up. "But . . . " she stammered—". . . I have the key!"

He brushed her words away with, "Some people don't need a key to start a simple little car like yours."

Shocked, Sarah said, "No one *I* know—"

Smiling wryly, Kight interrupted: "But, then, you've led a very sheltered life, haven't you! Today's incident alone proves that."

Sarah felt a flash of outrage. "What's that supposed to mean? I wasn't frightened for *myself,"* she protested, only half-truthfully. "It was . . . " Here she gestured toward Iris, listening wide-eyed on the nearby couch. "You've just *admitted* you know nothing about . . . so why not suspend "judgment" until you have some facts! For a change!"

He raised one dark eyebrow sardonically. "I keep forgetting about that red hair, although I can't think why. It's obvious enough."

Sarah remained silent while he poured a tall glass of

orange juice and handed it to Iris, then poured a finger of brandy and insisted that Sarah take it. "You both need a pick-me-up, Sarah, so for once will you just do as you're told, without an argument?"

Sarah reluctantly took a sip of the fiery liquid, wondering why Kight was so irritable. The way he behaved, one would think Sarah'd planned the whole nasty incident just to annoy him! He needn't think she couldn't have coped on her own. After all, it was his striking that young man that had made Iris faint, so sensitive was she to violence. Granted, Sarah had been frightened by the man's advances, but she hadn't been so "sheltered" that she hadn't dealt with unwelcome advances from men on the street before! Sarah was about to open her mouth and tell Kight all this, in no uncertain terms, when a wave of weariness washed over her. Let him think what he liked about her. What difference did it make, anyway?

She attended to what Kight was saying to Iris, some soothing nonsense to coax a smile and take the pinched look off her pale little face. "And as soon as I'm not so busy, remember that you and I are going to Coloma, where James Marshall first discovered gold. You'll like that, won't you?"

Iris nodded enthusiastically, her gamin smile nearly back to normal. "Sarah's coming, too, isn't she?"

Glancing briefly at Sarah, Kight said casually, "She can tag along, I suppose, if she cares to."

"Will you promise something else?" Iris asked, pressing her advantage like a seasoned opportunist.

With playful caution, Kight hedged, "I'll have to hear what, first."

"Promise that when I grow up you'll hire me to build your buildings, like those ladies we saw today."

First glancing briefly at Sarah, Kight answered Iris seriously. "When you grow up, you may do anything you're qualified to do. If you want to build my buildings, and you can handle the hard work—why, you're hired."

"Will wonders never cease?" Sarah said blandly. "May I ask just exactly when you hired your first woman construction worker?"

In a less than forthcoming manner, Kight mumbled, "Right after the legislature passed the Equal Rights Amendment, but—"

"Aha!" Sarah pounced. "You had to be forced by law!"

"As I was saying," Kight continued, "*but* since . . . January, I believe it was, we've gone out of our way to hire a good many more women."

Sarah and Kight exchanged a long, measuring look. At last, Sarah said thoughtfully, "I see."

Kight replied, "Do you? I hope so."

The elegant lines of the sumptuous Lincoln Continental looked totally out of place parked on the shady, tree-lined street in front of the Reillys' bungalow. At the sight of the car, and of its handsome, elegant driver, Maggie Reilly was struck dumb. It was Mike who did the honors, shaking hands with Kight in a dignified man-to-man way.

"We're honored to meet you, Mr. Ramsey. Do come in and have some refreshment."

But Kight refused with thanks, saying that he and Sarah were going to an early dinner. On hearing this,

Maggie gave Sarah an oblique smile, raising her eyebrows significantly. Sarah shrugged and, in dumb show, indicated that it was all news to her.

Iris poured out the story of the eventful shopping trip, and as she got to the more exciting parts, Maggie grew more and more agitated. "You see?" she interjected shrilly. "The streets aren't safe for decent people anymore."

"Oh, pooh, woman!" Mike said quickly, glancing worriedly at Iris' listening face. "I can well remember, in the years of your prime, when you couldn't walk a block without a dozen men giving you the eye. It happens to all pretty women—not to say it's appreciated! But no harm was done, so let's drop it."

"No harm was done because Mr. Ramsey happened along," Maggie murmured stubbornly, then shut up at the look Mike arrowed her way.

As Kight and Sarah were leaving, Iris suddenly cried out, "My new clothes are in your car, Sarah!"

Kight tousled Iris' sandy hair. "Don't worry, I'll deliver them to you tomorrow on my way to work."

"Can't you come when I'm here?" Iris asked in a pretty, coaxing way that Sarah'd never seen her use before.

Kight laughed. "All right, little minx. I'll come when you're home from school—if you aren't afraid you'll tire of the sight of me," he teased.

With a shy and demure smile, Iris replied, "I would never get tired of you, Kight."

"Nor I of you, Iris," he said, kissing her good-bye on her flushed cheek. Watching them, a half-smile played about Sarah's mouth. Really, how charming he was—with children. What a wonderful father he'd make for some lucky child; like Bonnie—poor, unhappy, insecure little Bonnie.

Kight and Sarah exchanged very little conversation during the drive to the early dinner he'd so confidently arranged. Only once did he say anything of consequence. While they waited for a red light to change, he said without looking at Sarah, "Bill Blanding won't mind that I'm taking you to dinner, I hope?"

Unused to living a deception, Sarah swiftly searched for an appropriate response. "I think not, under the circumstances," she said finally, meaning the afternoon's events and Kight's gallant rescue.

"What circumstances?" he asked quickly.

Having already thanked him for his help, and not wanting to clarify her remark for fear he'd think she was gushing, Sarah answered shortly, "I can still go where I like, I hope. I'm not married yet."

After the slightest hesitation, he said smoothly, "Just so. Nor am I—yet."

"Well, then," Sarah said grumpily, looking out at the after-work traffic, wending its way home in the dusk.

But the atmosphere in the little restaurant Kight had chosen was too seductive to resist, and Sarah soon succumbed to its romantically low lighting, charming country decor, and tantalizing food aromas. When the waiter greeted Kight familiarly, the image of him here with other women, with Vivica, did impinge on Sarah's pleasure, but she resolutely repressed the image, determined to seize the day and enjoy it, and let the future go hang. When Kight had given the solicitous waiter their order in his usual masterful, efficient way, and their cocktails had been served, Kight got right down to business.

"First of all, I want to know why Iris should have fainted today. I admit it was an unsettling experience, but I could see from where I was that you were handling it well enough—up to a point."

Ignoring his small dig, Sarah tried to explain to Kight, without seeming to blame him in the slightest, that it wasn't so much the young man's insolence that undid Iris as the sight of violence *per se.* However, to explain why *that* should be, of course Sarah had to tell him the whole sad story of Iris' short and, until a year ago, unhappy life.

They were well into after-dinner coffee by the time Sarah'd finished the story, and she suddenly realized she hadn't tasted a bit of what she'd eaten. She took a sip of water to soothe her parched throat and leaned back against the banquette with a dreary sigh. Kight slowly shook his head in a gesture of hopeless wonder at the baseness of what some members of the human race were capable of.

"What a rotten cur this Millidge is. When I think that some people would give anything—everything!—to have a child like that to love, and this scum treated her and her mother so vilely. . . ." His eyes held a dark and baleful look. "He must never be allowed to lay eyes on that child again," Kight stated with grim determination.

Again Sarah sighed. "No one disagrees with you there. But how to accomplish it under the law is something else again," she said tiredly.

Summoning the waiter and indicating that he wanted the check, Kight replied in a flat, no-nonsense tone, "There are ways."

Kight was silent on the drive back to the acreage, and Sarah had so much to think of herself that she barely noticed. For the first time since she'd known Iris, Sarah felt something like hope and confidence and peace in her heart on behalf of the orphaned child. If, through Sarah, this champion, Kight Ramsey, had joined Iris's cause, then all else Sarah had suffered by knowing him was well worth it. What if he was a bit of a philanderer? What if he had toyed with Sarah and ridden roughshod over her feelings? No one was perfect, after all. And any friend of Iris' was a friend of Sarah's.

"Kight?" Sarah said timidly, so deep was his concentrated silence.

"Humph?" he answered absently.

"I really appreciate what you are doing for Iris."

He took his eyes from the road and looked deeply into hers. His mouth relaxed, yet he didn't quite smile. "So we're to be friends at long last," he said softly.

"I'd like to, if you would," Sarah said shyly.

"All hatchets buried, then?" he asked. And when she nodded, he persisted with, "The future starts now, and no harking back?"

Sarah laughed. "Yes, if you like. No harking back."

Suddenly Kight flipped on the turn indicator and swung the car over to the shoulder of the highway.

"What is it?" Sarah asked anxiously. "Do you have a flat?" She peered at his face, dimly perceived in the light from the car's dashboard, and saw that he was smiling.

"A *rapprochement* of such magnitude must be celebrated somehow." Kight reached out to grasp Sarah firmly by the shoulders, kissed her soundly on the mouth, then held her away so he could look directly into her eyes. "Now it's done and sealed with a kiss. You can't change your mind anymore."

Sarah lowered her eyes so the tears forming in them wouldn't be seen. With a fervency she felt but dared not show, she answered, "I won't change my mind, I give you my word."

Kight nodded gravely, then took Sarah's hand as if it were a piece of delicate china and kissed her palm lightly.

When the darkened car again was purring through the cool, spring night, Sarah smiled at the fresh memory of Kight's mouth on hers, his kiss sealing their friendship. She smiled, though her heart was sore at the gross inadequacy of the word to describe what she now knew were her real feelings for him. She loved Kight Ramsey with all her heart, and she knew she always would. She also knew, now, this night, that she would never have the family she so much wanted. Because if she couldn't have Kight—and she couldn't, except as a friend—no one else would ever do. Sarah would remain a single lady, just as Aunt Elaine had done. But at least she had something to remember—just as she'd wished to have that first day when she'd chided herself for evading the touch of Kight's hand on her face. She had that now, a few kisses, and the promise of friendship. That would have to suffice.

So the momentous bargain so important to Sarah was struck and sealed with a kiss; and the coming weeks might have been very different, Sarah was to think later, if only the first thing she saw when they'd reached the parking pad at the acreage hadn't been a silver

Jaguar with a silver-haired Vivica standing beside it, her arms angrily akimbo.

When they stepped out of Kight's car, Vivica advanced on them, her beautiful face contorted with fury. For the first time Sarah knew what it meant to say that someone was beside herself with rage. Sarah was awed to see how perfectly Vivica's daughter, Bonnie, had learned to mimic her mother's epic tantrums.

"You had a date with me!" she shrieked at a stiff-faced Kight.

"Something came up—" he began in a frigid, calm tone.

"I can *imagine* what came up!" Her voice rose to the shrill volume of a band saw. "How *dare* you keep me waiting while you amuse yourself with a shabby little servant!"

Sarah blanched, but, nonetheless, waded into the fray. "Vivica, listen, it wasn't Kight's fault—I can explain!"

Then Kight turned on Sarah with what seemed like a loathing to match Vivica's and shouted, "That's enough! There's no explanation required here, and if there were, I wouldn't need you to make it for me. Go on to your own house now," he commanded.

When Sarah hesitated, again he shouted at her.

"Damn it, do as I say! Do you want to make matters even worse than they are?"

Sarah gasped at the unexpected assault, then turned to run like a life-threatened deer to the safety of the poolhouse. Once there she threw herself on her bed and gave way to heartbroken sobs, until at last she fell into an exhausted sleep.

Chapter Nine

The next morning Sarah awoke with a pounding headache and a general feeling that she was coming down with the Black Death. She got out of bed only long enough to call Mrs. Mole on the house phone and ask her to relay the message to Mrs. Ramsey that she wouldn't be working outside today. Sarah dozed fitfully most of the morning, and it was nearly noon when she heard a thump on her front door, followed by Mrs. Mole's truculent voice.

"Yoo-hoo, where are you, girl?"

"In here," Sarah called out weakly.

Mrs. Mole stood in the bedroom doorway with a long narrow parcel in her arms, and so floridly alive and energetic did she look that the very sight of her exhausted Sarah.

"See here what came for you," she said, thrusting the parcel at Sarah. With fingers like water, Sarah fumbled at the white ribbon, wondering without much interest who'd be sending her a present. Had she by any chance forgotten her own birthday? But when she saw the contents of the box, she knew well enough who'd sent it, although there was no card.

The box held one quite perfect red rose, picked at the precisely correct time, when the outer petal layer has just begun to open. Sarah laid the cool blossom against her cheek and closed her eyes, suddenly full of stinging tears.

Mrs. Mole smiled archly. "Is that from a gentleman friend?"

At the phrase, some dam burst in Sarah and she began to weep in earnest. No, not a "gentleman friend"—he would never be that—but at least a *friend*, that precious commodity she'd so newly gained and then had feared she'd lost forever.

"Sarah!" Mrs. Mole exclaimed and, as if censure were second nature to her, immediately began to scold. "Here now, shame on you! A big strong girl like you!" But at the same time she scolded, she also sat down on the side of the bed and lay her roughened hand gently on Sarah's forehead and cheeks to feel for fever. Then, finding none, she stroked back the tangled red hair from Sarah's forehead. "There, there, now," she crooned, "aren't you a silly goose to cry over a lovely present like that?"

Sarah gulped convulsively and nodded. Mrs. Mole murmured, "You're just upset over that ruckus last evening," causing Sarah to marvel at the speed and efficiency of the grapevine in this small community. "You don't want to take Vivica's temper too much to heart, Sarah," Mrs. Mole said confidentially. "She flies off the handle at most any little thing."

Sarah gave Mrs. Mole a small, watery smile of gratitude for making the effort to comfort her at the divine Vivica's expense. It was a handsome gesture, considering the source.

Reverting to her usual self, Mrs. Mole sternly ordered Sarah to stay where she was while that good woman prepared a bowl of hot vegetable soup for Sarah's lunch. Over Sarah's protests that she mustn't bother, she said with all the authority of a prison matron, "I'm going to fix it, and you're going to eat it; so save your breath to cool your soup."

Sarah sank back onto her pillows feeling like a reprieved criminal. She lifted the perfect long-stemmed rose to her lips, knowing that it was meant as an apology for Kight's outburst last evening, but nothing more. His words, though ill put and ill timed, had been clear enough. "Don't interfere between us. Don't make things worse than you already have." In the future Sarah would take scrupulous care never again to encroach on Kight's life.

A week later on a day too cold and wet to work outside, Sarah, knowing that Mrs. Ramsey's housewarming party was only days away, offered her services in the house. So it was that the two women sat at the newspaper-covered dining room table, companionably polishing flatware. An apricot-wood fire crackling in the fireplace made the already comfortable room even more intimate on such a dreary day.

"Sarah, dear, what will you wear Saturday night?" Mrs. Ramsey asked.

"Saturday?" Sarah repeated, confused.

"To my party—my housewarming party."

"But I thought it was Sunday afternoon," Sarah said.

Mrs. Ramsey looked up from her butter knife. "It's never been planned for a Sunday, dear. Where did you get that idea?"

"Why, Vivica mentioned it, I think," Sarah said slowly, thinking back to that day by the pool.

"I see," Mrs. Ramsey murmured. "She must have got it mixed up somehow. She leads such a hectic life, she often muddles things. Good thing I asked, isn't it? Or you would've missed the party."

Sarah hesitated, then plunged in. "As to that, Mrs. Ramsey, it's awfully nice of you to include me, but I really don't think—"

"Nonsense," the woman said briskly. "You must come—I insist. Now we'll say no more about it, shall we?"

Defeated, Sarah nodded. In the four months she'd worked for Grace Ramsey, she'd learned that this kind and charming woman also had a will of iron, and to oppose her beyond a certain point was utterly fruitless.

Abruptly changing the subject, Grace Ramsey said, "Vivica tells me you're planting pyracantha on that slope by the pool."

Sarah hesitated, then replied evasively, "The last time it was discussed, that's where it was left."

Mrs. Ramsey nodded as if it were of no great moment, then continued with what at first seemed to Sarah like a non sequitur, although she should have known better.

"Vivica thinks of Kight and me as her family, you know, Sarah—one might say her adopted family. She has done so since she was quite a young child. Her mother has always been such a busy, socially involved woman, and of course her father was gone quite a lot while he and Kight's father were building the business. So one might say that Vivica gravitated to the next best thing.

"Because she thinks of herself as a family member, she has a tendency to overstep her authority from time to time. As I think I've mentioned before, Vivica is a very impulsive young woman. Soon enough she'll have her own household to manage, but until then I wouldn't want to burden her with the management of mine." Grace Ramsey peered up from her busy hands to see if Sarah was listening. Sarah, hanging on the older woman's every word, nodded.

With a small smile, Mrs. Ramsey continued. "Now you, on the other hand, did not gravitate here but were very carefully hand-picked. You were picked for your skill, your knowledge, and your judgment. Isn't that so?"

"I hope so," Sarah murmured.

"Well, then," Mrs. Ramsey inquired, "what are you going to plant on that slope?"

So there it all was, Sarah thought, spelled out as clear as the nose on one's face—Vivica would soon have the right to give orders to the help, but not quite yet. Sarah smiled with relief and felt an urge to reach across the table to hug this remarkable woman.

"Xylosma, I thought. It has such fresh, shiny leaves and it's such a good ground cover. And it doesn't bloom, so bees will never be a problem by the pool."

"Excellent," Mrs. Ramsey said comfortably. "Just what I'd have chosen myself."

The guests' attire at Mrs. Ramsey's party that Saturday night was the usual California conglomeration. Californians, true to their relaxed and various life-styles, wore what guests in the more staid and traditional parts of the country would only have worn to a come-as-you-are party. There were gray-and-white pin-striped suits, jeans and chiffon tops, disco dresses, long and short dinner dresses, tuxedos, sport jackets, and Hawaiian muumuus.

Sarah wore, at long last, the silk Chinese dress that Ben Yashimoto had given her so many years before. She knew that its close fit suited her height, as its peach color complemented her auburn hair. And although she knew she looked well enough, she felt quite awful.

In spite of the gaiety of the party's mood, the charm of the newly decorated house, and the interesting and delicious food, Sarah found it impossible to enjoy herself. Everywhere she looked, it seemed, a painful image in black and white haunted her eyes. Kight, handsome at any time, took Sarah's breath away tonight. Never had a man looked more elegantly sophisticated than he in his superbly cut black tuxedo and ruffled white shirt. Vivica, flawlessly beautiful at any time, wore a white satin evening dress that made her seem a mythic ice maiden. But what caused Sarah such pain was the sight of Vivica clinging so familiarly and closely to Kight's side that she might have been another layer of his skin.

Once, during the few seconds that Kight left Vivica's side to get her another drink, she sauntered over to Sarah, standing alone for the moment, and smiled patronizingly.

"How nice you look, Miss Halston, although I'd expected to see you in your new dress."

Sarah murmured coolly, "It didn't seem quite right for the occasion."

"I see," Vivica said with a smug smile. "But how fortunate that you had something else you felt was right. Or did you buy it especially for the party?" she asked brightly.

When Sarah made no reply, Vivica said, "How brave you are to come to a party of strangers all by yourself!" Looking around vaguely, she added with false concern, "I do hope someone has introduced you around."

When Sarah merely smiled briefly in reply, Vivica lowered her voice and said confidentially, "I believe I owe you an apology, Miss Halston. I behaved in a truly ridiculous fashion last week, I admit it. As if a serious working girl like you would even contemplate any sort of personal relationship with a man like Kight! Or he with you! I can't think why I ever thought . . . Well! I was probably upset about something else—I can't even remember what it was. I am sorry, and I know you'll forgive me. Anyway, no harm's done." She smiled brightly. "As you've no doubt noticed, Kight and I have made up. He's such a dear, you know. So indulgent of my silly little flare-ups. But then, he does like a woman with spirit," she said roughishly. And moving off, she waved a good-bye to Sarah with delicate pink fingertips.

With an aching heart, Sarah left the cheerful noise and the perfumed air of the house behind her and stepped out into the garden—where she belonged. But even the garden was hers only on borrowed time. Tonight, seeing Vivica and Kight as close as hand in glove, had put to death what little was left of Sarah's silly, romantic, and impossible dreams. When the fête in June was over, when her obligation to Mrs. Ramsey was satisfied, Sarah would resign her position here and go—elsewhere; anywhere would do so long as she didn't daily have to live with the pain of seeing Kight belong to someone else.

With a slow and heavy step, Sarah passed the rose garden on her way to the little white gazebo she'd had built a few yards away. Today, in mid-April, the first rose had opened, a stunning red rose named Mirandy. Sarah stopped to pass her fingertips over its velvet face and stooped to inhale its sweet aroma. From behind her came Kight's deep, quiet voice.

"Now, that's what I call a beautiful rose."

Accustomed by now to having Kight turn up at unlikely times and places, Sarah turned around calmly to see his white shirt glowing in the soft, fragrant dark.

"The first rose of summer," Sarah said quietly.

Kight bent his dark head to smell the blossom, then said thoughtfully, "Quite a contrast to a florist's rose: so chilled from refrigeration the aroma's killed, the petals so perfect they might be porcelain or silk. But *this* rose smells like heaven, its petals are still warm from the sun, and here's a tiny flaw where an insect made a noonday meal, perhaps. It's like the difference between life and art, isn't it?"

Sarah smiled into the dark, her heart made a little lighter by his words. "Still, I thank you for the florist's rose. It was very kind of you."

Suddenly Kight grasped her upper arms in his hands and gave her a little shake. "Oh, Sarah, I'd rather have given you *this* one, if only it had opened in time. And it wasn't *kind* at all!"

Then his warm, firm lips came down on hers and his arms encircled her body and pulled her close. Thoughts of resisting him swirled in Sarah's mind. How bittersweet it was to be held by him this way, to taste his kiss, knowing there would never be anything more than this and knowing that even this was stolen from its rightful owner. Before the irresistible pull that pulsed in her heating blood could weaken her will beyond all resistance, before she surrendered to the flame of wild desire that she felt for him and that his lean, hard body told her he felt for her as well, she wrenched away from his embrace, although it tore her heart to do so.

In the dim light of the moon shining on his dark and agitated face, Sarah saw the passionate look in his eyes and knew she'd been right to stop him before something happened that both of them would sorely regret. He looked away abruptly, and in a strained and muffled voice he said, "You're right, you're right—this is not how it should be."

"It never will be, Kight," Sarah said dully. "You know that as well as I do."

He stared at her as if puzzled. "But, Sarah, our bargain . . . I thought . . ."

"To be friends, I believe it was. Nothing more. Or is your definition of 'friend' so different from mine?"

A glowering, ruthless expression overcame his features and the corners of his mouth stiffened. He took a step toward her and a cold fright washed over Sarah. In a pained and frantic voice, she exclaimed, "There are other people to consider! Maybe you can forget that," she cried, "but I can't!" Curse the man for making *her* protect Vivica's interests!

As if her words had been a pail of cold water dashed in his face, Kight stopped dead in his forward movement toward her. His face closed up like a stranger's and his voice was distantly courteous as he said, "I've been very stupid and I hope you'll forgive me. I give you my word this will never happen again. May I see you back to the house now?"

With a weeping heart, Sarah allowed him to take her elbow and lead her toward the house, just as he had that first day she'd met him. Oh, if only that fateful day had never taken place! If only Sarah could go back to the girl she'd been, with no one to love and a heart that wasn't shattered in a million pieces.

With a chilly, polite bow, Kight took his leave of her in the foyer, and without a backward look, he plunged back into the noisy, laughing crowd. Sarah took a deep, shuddering breath and looked around for Mrs. Ramsey. She would find her hostess, pay her thanks, then take her leave. It was nearly midnight, anyway, and no one here would miss her company.

After a search of the house, Sarah found Mrs. Ramsey in Kight's study speaking on the telephone. When she saw Sarah at the door, she said into the phone, "Oh, here she is now, Mr. Blanding. Hang on."

Sarah took the phone with all the dread that any unexpected phone call late at night brings. "Bill, what is it? Is Iris all right?"

"Yes, she's perfectly fine, Sarah. But Millidge is back in town. He called me a few minutes ago—drunk, of course. Says he's come to take back his daughter who the law 'stole' from him." Bill's voice deepened in sarcasm.

Sarah gasped and her eyes involuntarily sought those of Grace Ramsey, who stood nearby listening unashamedly.

"What is it, dear?" she asked anxiously.

Bill continued, telling Sarah that Jack Millidge couldn't find Iris' whereabouts at least until tomorrow, when the Friends of Childhood office opened.

"But surely no one there will tell him where she is!" Sarah cried in disbelief. "They all know what he's like!"

Mrs. Ramsey muttered worriedly, "I'll get Kight." And before Sarah could stop her, she'd hurried from the room.

Bill replied, "They're obliged by *law* to let him see his daughter, Sarah. You *know* that. God knows you've complained about it enough." But she wasn't to worry, he said. He'd arrange for Jack Millidge to see his daughter at the Friends office, not at the Reillys' home. He was sure the judge would go along with him on this bending of the law, under the circumstances. And Bill would also put off the interview until Millidge had a chance to cool down and come to his senses. And sober up.

Disgusted, Sarah raised her voice to nearly a shout. "Oh, can't you forget about your precious due process for once? And what about yourself? You're in danger, too! You ought to get out of your apartment right now, go to a hotel—"

"You forget my address isn't in the phone book," Bill replied calmly, "and for just this sort of reason. He can't find me any more than he can find Iris. So now you see there's really nothing to worry about. I only called to tell you because I was afraid if you found out from someone else, you'd never forgive me. But if you're going to make yourself sick with worry, I'll be sorry I told you."

The hurt in Bill's voice squelched Sarah's temper and she sighed softly into the phone. "No, Bill, you must never be sorry on my account. I know I've behaved badly, and I'm sorry. Of course I'm glad you told me, but you could never do anything I wouldn't forgive you for, Bill. How could you even think it, knowing how I feel about you?"

Sarah heard a slight sound and looked up to see Kight in the doorway with his mother behind him. Had he overheard enough of what she'd said to realize what was going on? she wondered. If only she could ask for his help now that Iris was in real danger, but she'd vowed that never again would she encroach on his private life. And that incident in the garden was a blazing example of where it would lead if ever she did so again. While these thoughts were spinning through her mind, Sarah was also listening to Bill on the phone. He was telling her that she wasn't to worry—fat chance of that!—that he would keep her posted, and that she was to spend her day with Iris tomorrow as usual. Sarah answered no, she wouldn't; yes, she must; and yes, of course she would. Then she said good-bye and hung up the phone.

Kight reached behind him and closed the door to the study. "May I know what the trouble is?" he asked courteously.

"It's a personal matter," Sarah replied, "nothing you need trouble yourself with."

Mrs. Ramsey broke in: "Oh, but Sarah, dear, you looked so horrified. If you're in trouble, please tell us—"

There came a sharp rap on the door, and Sarah

heard, in muffled but audible tones, Vivica's honeyed voice outside the door. "Kight, darling? Are you in there?"

Kight turned to lock the door, then called to her briefly, "I'm busy, Viv. Run along now. I'll be with you shortly." Then, to Sarah, he repeated his mother's words. "If we can help in any way—"

"Thank you both very much, you're very kind, but really, it's nothing—just a personal matter, really—" And she began to edge toward the door.

Kight, his eyes narrowed speculatively, stepped into her path and looked down at her searchingly. "You haven't forgotten that we're to take Iris to Coloma tomorrow, have you?"

Sarah frowned. "Perhaps we could put it off until another day. I don't think I feel quite up to it. I'll keep her here with me—"

"There are other people to consider, Sarah. Maybe you can forget that, but I can't," Kight said dryly, quoting her exact words of a few minutes ago in the garden. "Or perhaps your definition of friendship is different from mine, but I promised my little friend an outing and I have no intention of disappointing her. So will you stay home, or come along with Iris and me?"

What a beast he was to plague her now, of all times! And so unfairly, at that. Even though Sarah realized that once again Kight was blackmailing her through Iris, she also realized that Iris would be safer with Kight and herself in Coloma than she'd be at the Reillys' in Sacramento.

"All right, I'll come along," she agreed. "I'll fetch Iris early tomorrow morning and you may call for us at the poolhouse at your convenience."

As if she'd never spoken, Kight said coolly, "You and I will fetch Iris early in the morning, in my car, and then we'll go to Coloma directly from the city. Be ready by eight o'clock."

Too exhausted and worried to argue further with him over something that hardly mattered, she agreed. With a reassuring smile to Mrs. Ramsey that couldn't have been very convincing, Sarah left the house through the kitchen to avoid the party, which was still going strong.

When Kight and Sarah arrived at the Reillys' house the next morning a little after eight o'clock, she was astonished to see not only Iris waiting excitedly in the living room, but also Mike Reilly, with his battered toolbox, and Maggie, clutching her worn black purse. When the two women were settled in the back seat of the roomy Lincoln, with Iris in the middle up front, Maggie whispered to Sarah, "Mrs. Ramsey called at seven this morning, apologized for giving us such short notice, and asked if Mike and me could come give her a hand clearing up the house after that big party she gave last night. She seemed like an awfully nice lady, Sarah. Is she?"

Stunned, Sarah could only nod. And it wasn't until the Reillys had been delivered to the acreage and introduced to Grace Ramsey and Agnes Mole, and the three travelers were back on the highway, that Sarah had a chance to broach the subject.

"I was very surprised that the Reillys were coming to the acreage this morning," she began.

"Oh, yes?" Kight replied unhelpfully.

"I suppose you told your mother about meeting them?"

"That's right," he said laconically.

"I see. I'd have thought Mrs. Ramsey would've mentioned it to me if she'd planned to hire them as occasional day help."

"I suppose she didn't think she needed your permission," Kight said blandly.

Sarah flushed with anger and flared at him indignantly, "Of course not! You know I didn't mean it that way! What a beastly rude thing to say!"

Kight glanced at her briefly over Iris' head, then turned his attention back to the road. "I'm sorry. You're right, of course. Mother would've mentioned it, except that it came up so suddenly. When the guests finally left last night, she realized there was more clearing up to do than she and Agnes could handle. So I suggested she call the Reillys since I was going there in the morning, anyway. It all worked out very well. And surely you're glad"—he turned to Sarah with a questioning expression—"that your friends will spend a peaceful day in the country."

Sarah was silent, sifting his words for any hidden meanings, and finding none, she smiled. "Yes, I can't think of anything nicer than for them to spend today in the country."

Reassured that three of her loved ones were safe at least for the day, Sarah gave her thoughts to the day's outing. She pointed out to Iris the change in the landscape as they climbed ever higher from the suburbs and exurbs of the city into the beautiful foothills country of the noble snow-capped Sierra beyond. Domesticated shrubs and flat lawns gave way to scrub growth and the smooth rolling knolls, now green, but soon to be turned a burnished gold by California's coming rainless summer. The earth was mineral-rich here, and red instead of brown, and the always wide and vivid blue sky broadened overhead even more until it seemed they were encompassed in a world of blue.

"My ears are stuffed," Iris announced.

"It's the altitude, dear; just swallow and they'll clear," Sarah said.

As they climbed even higher, great stands of valley oak, manzanita, madrone, and incense cedar stood sentinel as they'd stood for thousands of years on either side of the gently winding highway that had been carved through the rocky hills. When they reached Coloma it seemed a forest itself, surrounding a small cleared settlement. The pridefully kept picnic and camp grounds were cool and quiet under their canopies of ancient, massive trees.

"Look, Kight! That must be the mill where gold was discovered!" Iris exclaimed excitedly.

"So it is," Kight replied, pulling the car into the large parking area just yards away from where the sawmill stood. There was a group of visitors clustered around, watching the great wheel in operation, and listening to a handsome young park ranger relate the story of those great and gaudy days. The three of them joined the group to listen.

"What you see here, folks," the ranger lectured, "is a replica of the original mill. It's ironic, but sadly true, that Sutter and Marshall's mill, like the rest of their business enterprises, fell into decline and ruin after the discovery of gold, for which they were responsible. When the hordes of miners flocked into the foothills in search of gold, the mill fell into disuse and was eventually washed away by floods in 1862. In 1924 some of the original timbers were found buried in flood debris, and we have them now in a safe shelter where you can see them, across the road. But this mill is as faithful a copy as possible, made with hand-adzed timbers fastened with oak pins, just like the original."

Sarah whispered to Kight, "The mill isn't very big to have changed the destiny of a whole state, is it?"

"Only sixty feet long, thirty-nine feet high, and twenty feet wide. And it not only changed California's destiny, but the entire country's. Some historians think that without the profits from gold that were invested in the North, the Civil War would have been lost to the South."

Sarah said admiringly, "You're as good as a private guide."

Kight shrugged. "The gold rush has always been an interest of mine. If you've seen enough here for now, let me show you the museum."

As they returned to the car, Sarah's heart was warm to think that, because of Kight, this day that might have been fraught with worry and tension was going to be a lark.

In the place of honor, in the middle of the small museum's main room, was an old stagecoach that looked as if it had seen many a hard trip across the miles. Made of wood, it stood high on its iron wheels, with an interior smaller than most modern cars, but intended to seat six people—six uncomfortable people, Sarah thought. Attached to its door was a list of regulations for stage passengers which Kight read, in part, to Iris.

" 'The best seat inside the coach is next to the driver.

Even if you get sick riding backward, you'll get over it and will get less jolts and jostling.'

" 'When the driver asks you to get off and walk, do so without grumbling. If the team runs away, sit still and take your chances.'

" 'In cold weather abstain entirely from liquor. You will freeze twice as quickly under its influences.'

" 'Spit on the leeward side.'

" 'If you have anything to drink, pass it around.'

" 'Never shoot on the road, as the noise might frighten the horses.'

" 'Don't discuss politics or religion.'

" 'Don't point out where murders have taken place, especially if there are women passengers.'

" 'Don't grease your hair; travel is dusty.'

" 'Don't imagine for a moment that you are going on a picnic. Expect annoyances, discomfort, and some hardships.' "

In spite of these warnings, Iris said wistfully, "I wish I could have a ride in it."

"So do I, honey," Kight replied. "But I'm afraid the old buggy's riding days are over. Come over here and I'll show you Marshall's first flake of gold."

In the middle of a showcase, displayed on a small pillow of red velvet, sat the small, insignificant flake of gold that had started the massive adventure of 1849. It reposed there surrounded by chunks of some forty other mineral samples found in this rich soil, such as malachite, cinnabar, mica, and opalite.

"But it's so small!" Iris said. "I thought it would be big."

Kight laughed and tousled Iris' hair. "You'd be just

the type who passed wild stories around during the rush. There's one story that a miner found a nugget weighing 839 pounds. He was so afraid to leave it for fear it'd be stolen that they say he sat on it for days, offering as much as $27,000 to passersby for a plate of pork and beans."

After Iris had scrutinized the museum's every exhibit, Kight took them to the charming, old Nevada House Hotel for lunch. Iris was enchanted by the rugged decor: a mammoth bellows hanging on the wall, an oxen yoke and bow hanging from the ceiling, and a tall, heavy, gleaming old brass scale for weighing gold, securely locked up in a wall case. The walls and carpet were a chocolate-brown, which set off the bright red table linen and white crockery dishes. Sarah settled back happily in her chair and left the ordering to Iris and Kight.

An hour later, Iris was sound asleep in the back seat of the cool, quiet car; Kight seemed intent on his driving; and with every mile the powerful car consumed between here and home, Sarah's recent good spirits dissipated. Without Sarah's awareness, a heavy, troubled sigh escaped her.

The feel of Kight's hand on her own brought her back to herself. She looked questioningly at him, while gently pulling her hand away. When he turned to meet her eyes, there was something in his face that alerted her to trouble.

"Sarah, there's something I have to tell you," he said formally. "When we get home, my mother will have told the Reillys what you know, what I know, what we all know now, except Iris."

"What!" Sarah exclaimed. And she was immediately

hushed by Kight's jerk of the head toward the sleeping child. "But how do you know?" she asked in a low, intense voice.

Kight's face took on a remarkably sheepish look, and this alone threw Sarah even further off stride. Imagine what it must take to make the arrogant Kight Ramsey feel sheepish!

"I took the liberty of calling Bill back last night after you'd left," he said defensively.

Sarah gasped. "But I told you it was a personal matter! How dare you interfere—"

He made a weary gesture with his hand. "If it makes you feel any better, I did give it long thought before I called him, in case it *was* just a personal matter between the two of you. And if my mother hadn't been so sure it was more than that, I assure you I wouldn't have interfered. But I had a hunch it was something to do with Iris, so I took the chance and called. Bill told me about Millidge, and I could see that the Reillys were sitting ducks. I got them out of there by having my mother call and ask for their help. I didn't dare tell them the truth for fear Maggie would panic and Mike would feel he had to stay put and play hero. And I didn't tell you because . . . well, I was afraid you'd react just as you have reacted."

Sarah scarcely knew what to think, or feel, let alone what to say. How like him it was to plunge in and take charge. But bless him for it this time! she thought with great gratitude. "I don't know how I'll ever thank you, Kight, for everything you've done today. I'll be forever in your debt."

"There is something you can do for me, actually," Kight said, looking at her from the corner of his eye.

Sarah said fervently, "Just tell me what it is, and I'll do my best to oblige."

"Bill told me quite a lot about Iris's situation last night—that she can't ever be adopted until her father gives his permission."

"Which he'll never give," Sarah added bitterly.

"Yes, Bill says that, too. He says that Millidge knows that's the only power he has left—to obstruct her future happiness, even if he doesn't want her himself. But *I* say he can be *persuaded*," Kight said ominously. "And if I can persuade him to relinquish her, *you*, Sarah, could adopt her and she'd be safe."

Tears sprang to Sarah's eyes, for Kight had voiced the dream she herself had dreamed a million times. But he didn't know the full extent of the obstacles in the way of that dream's ever coming true.

"I'd give everything I own, now and in the future, if only that could happen, Kight. But I would have a tough time persuading the court . . . my age . . . no steady job . . . She sighed. "No," she continued sorrowfully, they probably wouldn't let me adopt her. But someone else could—if you could make her father sign the papers." Excited by even this dim hope that Iris would someday be securely placed in a family of her own, Sarah turned shining eyes to Kight.

"It would be so wonderful if you could do that."

"But if you were married?" Kight asked.

Unsure of his meaning, Sarah hesitated. "But I'm not."

"You said you'd do your best to oblige me, Sarah. I'm asking you to marry me. Will you do it?"

Chapter Ten

"Of course not," Sarah said coldly.

In real time, the answer came only seconds after the question. But as the heart measures time, Sarah agonized long and hard over the answer. How cruelly unfair it was to be offered her heart's desire, but under such circumstances that she couldn't, with honor, accept it.

Yes! Yes, she wanted to marry Kight, to see him daily, to live with him in the same house, sleep with him in the same bed, to bear his children—yes, if he'd loved and wanted her. But it was for Iris's sake that he asked, and God bless him for it. And even then, *yes,* she'd marry him for Iris's sake, except for one irrefutable fact: a marriage with love on one side only was doomed to eventual failure.

Sarah was fully aware of Iris' growing affection for and trust in Kight over these past months. If he became her father, the child would give him her heart totally. Then, how devastated she'd be when she lost him again as she inevitably must in a year or two, or five or seven. And when that happened, not only would Sarah's heart break, but, much worse still, Iris might never recover from the loss of him.

And for Kight's own sake, as well, Sarah must refuse what she most wanted in the world. He was on the verge of marrying a woman he *did* love, had loved for many years. Hadn't Agnes Mole said the two of them would have married long since but for Vivica's previous marriage? And then there was Bonnie, so sorely in need of a father, and with a prior claim to Iris' for Kight's love and protection.

And what if Sarah threw all honor and good sense to the winds and married Kight in the blind hope that somehow it would all work out? But there was no guarantee that Kight would persuade Jack Millidge to sign adoption papers, and if Kight failed—as so many had before him—he'd be saddled with a women he didn't love and deprived of the child for whom he'd made such a heavy sacrifice. And even if he were successful, there was no guarantee that their claim to Iris would prevail over families whose names were higher on the waiting list for adoptive children.

No, it was a hopeless idea. There was only one thing to recommend it: Sarah's deep love for Kight. But balanced against that one selfish desire were a thousand reasons against it—reasons she dare not mention for fear he'd sweep them aside, in his usual fashion, and weaken her already shaky resolve to do what was best for all concerned.

"Of course not," Sarah said coldly. "What a perfectly wild idea." Never had words come so near to choking their speaker.

After a short, bleak silence, Kight's mouth twisted

into a grimace that might have been intended as a wry smile. His voice was thick and ragged with some unpleasant emotion, probably hurt pride, Sarah thought, as he replied, "Well, nothing ventured, nothing gained. It seemed a good idea for a moment there."

"Oh, it was," Sarah said, more gently now, "and one I've thought of myself many times—but not with you." Then realizing how insulting that sounded, she amended lamely, "I mean, with someone else. . . ."

"You mean Bill." Kight's eyes were so intent on the road ahead that he might have been traveling hairpin mountain curves on a black, rainy night.

Sarah swallowed threatening tears. "Yes, of course. Bill."

"I thought of that, naturally, but since you two seem to have no immediate plans along those lines . . . or none that I know of . . . I thought perhaps you didn't . . . or he . . . Well. It's none of my business, as you've so rightly pointed out a number of times. All the same, to clear the way for you—for *someone*—I intend to do as I said: persuade that maniac father of hers to let that child go. The Reillys won't be allowed to keep her forever, and she must have a chance for a settled life with a family who will love her as she deserves."

When they reached the acreage, Sarah was confronted with even more bad news to add to her already full measure of misery. As they entered the foyer of the house they were met by an agitated contingent made up of the Reillys, Agnes Mole, and Grace Ramsey, all of whom must have been listening for the sound of the car with the acute hearing of a flock of bats.

As if by prearrangement, Agnes whisked Iris off to

the kitchen, promising cookies and milk, and as soon as she was out of hearing, Maggie Reilly burst into a flood of tears and everyone began to talk all at once.

What it eventually boiled down to was that Jack Millidge had lain in wait for the first arrival at the Friends office early that morning, and that first arrival had been a young receptionist, fairly new to the job. His wild, drunken ravings might have been sufficient to scare the girl into giving him the address he sought, but like the bully he was, he ensured his success by waving a switchblade knife in the terrified girl's face.

After he'd left with the Reillys' address, the stunned girl had gathered her wits enough to call Bill, who'd called the police. Millidge, further enraged at finding no one home, had little time to do more than break the front window, slash the living room furniture to shreds, and lay waste to every object in the house that wasn't nailed down, before Bill and the police arrived. Millidge was hauled off, shouting terrible threats, to the county jail, where he now resided for the time being.

Calmer now, Maggie spoke: "I don't care a pin about our belongings, so long as we're all safe. And that's all thanks to you, Mr. Ramsey," she said fervently. She turned to Grace Ramsey. "And you, too, ma'am, making it so easy for us to come here this morning—not letting us suspect a thing. God bless you both for what you did today."

Shuddering to think what might have befallen the Reillys this morning, Sarah, too, said a silent prayer of thanks for all that Kight had done. It was decided that Mike and Kight would return to the Reillys' house to salvage what they could from the wreckage, but for the time being, Mike, Maggie, and Iris would make the acreage their home.

During the weeks that followed, but for one deep shadow on her heart, Sarah was well pleased with life. The party was only days away, and the garden, even more of a showplace than Sarah had envisioned it that cold, rainy day six months ago, was ready. The spirea bushes, which now Sarah avoided calling by their common name, bridal wreath, were in ebullient, blowsy flower. The strawberry patch could have fed an army. Every bush in the rose garden was covered in blossoms, and in bloom, too, were the irises sheltered under the birch grove, and the primroses nestled around the gazebo. The lawn Sarah'd once despaired of now looked like the ideal meadow, its rich, lush grass sprinkled tastefully with the ferny leaves and bright orange blossoms of the wild California poppy and the cunning, miniature white flowers of the lawn daisy.

During this period of time, Sarah learned with amazement and joy that the charity for which she'd worked so hard to ready the garden was none other than the Friends of Childhood. As surprised as Sarah, Grace Ramsey had said, "I'm sorry, Sarah, dear, I've assumed you've always known! Self-centered of me, but I suppose I take it for granted that everyone close to me knows of my connection with Friends, since I was a founding member and now serve as chairman of the board."

But however the oversight occurred, it worked out well for all. Because Sarah pointed out Agnes Mole's love for children to Mrs. Ramsey, she was offered the position of housemother at the Friends shelter house. And Agnes, after a watchful period to determine if the Reillys were worthy of trust, opted to leave the Ramseys in their capable hands. She in turn commandeered the shelter house, saying she could wait there as well as at the acreage for Vivica to set up housekeeping again, and send for her. And besides, she forcefully pointed out, someone with sense was needed to look after those poor little tykes whose own folks hadn't the sense God gave a walnut.

So the Reillys were officially hired on at the acreage: Maggie as housekeeper and cook, and Mike as general factotum. The opportunity to be useful once again, as he'd been used to all his life, had mellowed Mike's irascibility beyond belief.

But certainly best of all, Jack Millidge was now a resident of Florida, an entire continent away from his daughter, although, under the law, Iris was no longer his daughter. Because he'd been so unwise as to give the law something to jail him for—mild though it was compared to what he'd done in the past—he'd forced himself into the position of having to "plea-bargain."

Before he departed to reside on the palmy beaches of Florida, he'd signed the paper that gave his permission for Iris to be adopted—and another paper, as well. This one acknowledged that he understood fully that his presence, no matter how brief, in the state of California, or any attempt to communicate with one Iris Millidge, a minor, without the express and written permission of her guardian under the law, one Sarah Halston, would result in his immediate incarceration to serve the jail sentence suspended herewith in return for his agreement to the aforesaid conditions.

"But however did you get him to surrender so completely?" Sarah asked Bill when he'd told her the joyful news.

Bill smiled with chagrin. "It wasn't me, Sarah, I'm sorry to say. It was Kight. Neither he nor Jack would admit a thing, but I think Kight made Jack an offer he couldn't refuse, if you know what I mean."

Uncertainly, Sarah asked, "You mean—a bribe?"

Bill shrugged. "Maybe. Or maybe partly, just *partly*, money. But mostly, I think, Kight offered Millidge the freedom from the constant fear of things that go bump in the night."

"Oh, I see," Sarah breathed with shivery awe.

"He's one man in a million, Sarah, that Kight."

Sarah lowered her eyes and gazed into a future now bright with promise for Iris, thanks to Kight Ramsey. "Yes, I know," she murmured.

Bill saw the vulnerable softness in Sarah's eyes.

"I thought you did," he said thoughtfully.

It was mid-June, only days before the great event, and Sarah fussed in the garden like a mad thing, seeing to every last detail so that the garden would do Grace Ramsey proud. Since Agnes had taken charge of the children at the shelter, Vivica was seldom seen at the acreage anymore, so it was with some surprise as well as dismay that Sarah heard the roar of the silver Jaguar racing up the smoothly paved driveway. Sarah clipped one more blown rose from a bush and straightened up to wipe the perspiration from her brow as she watched Vivica lithely swaying toward her, dressed in a smart pleated white skirt and sleeveless blouse.

"Hello, there, Sarah," she caroled. "Still slaving away, I see."

As she drew nearer, Sarah thought there was something different about Vivica's face, the muscles around her mouth less drawn, her lips fuller. Then suddenly Sarah realized that for the first time since she'd met Vivica, the other woman looked sincerely contented, even happy.

"I must say," Vivica said, looking perfunctorily around her, "you've done wonders with this place. I wouldn't have believed it possible. When Kight first brought me to see it, I told him, 'Darling, you can't be serious! The place is barely fit for peasants!' But now, just look at it. The house is presentable, at least—and the grounds are quite nice. What a sweet, rustic little weekend hideaway it will make."

Grateful that she'd soon see the last of Vivica, Sarah restricted her response to a polite smile. Then in a brightly curious voice, Vivica asked, "How does your little family of waifs like life in the country? Settling in all right, are they?"

With a sudden rush of bitterness, Sarah thought: *What a vixen the woman is!* "We're all very happy here," she muttered.

"I should think so!" Vivica wandered to a nearby bush and snapped off a rose bud, and at the painful sight Sarah suddenly thought of dear Ben and had to bite her tongue to keep from shouting at the future lady of the house to keep her destructive hands to herself.

Vivica turned suddenly, swinging her shoulder-length blonde hair fetchingly—although there was no one worthy to see the gesture—and said thoughtfully, "You know, the garden is really quite pleasant now. In fact, it's much too nice to waste on just a charity party. Think what a charming country wedding one could have here—for instance, right over there by that funny little gazebo, one could exchange the vows."

When Sarah, struck dumb by the tears frozen in her throat, didn't reply, Vivica turned to look at her.

"Don't you see it? Look here, the orchestra could play from over there—" She waved her arm vaguely toward the house, but Sarah had taken all she could bear, and with a wordless, mumbled sound, she began to walk away.

Vivica stood there looking after her in surprise, and in a mildly exasperated voice, she called after Sarah's receding back, "The trouble with you working people is you simply have no imagination!"

By three o'clock on the great day, Sarah had done everything conceivable to make the garden ready for the party, and now it was time to make herself and Iris ready. When Sarah took her newly purchased floral chiffon in peach and soft yellow from the closet, Iris protested, "But Sarah, I thought we'd be twins!"

"No, darling, you wear your dress from Kight, but I might be helping with the food, you know, and I'm afraid I'll get mine dirty."

Iris hesitated, then said mournfully, "You're going to hurt his feelings, Sarah."

With a brisk smile to hide the pang in her heart, Sarah said, "Oh, pooh, he'll never even notice." Someday, perhaps, years and years from now, Sarah'd have the heart to wear the Mexican wedding dress, even if Vivica had picked it out, but not now, not today; she couldn't have borne it.

A little later the two of them were made much of at the main house. Iris, a demure little shepherdess, perhaps by Gainsborough, in her white lade dress and a pink ribbon trailing down her sandy-colored hair, had the glow of true beauty on her newly tanned and relaxed little face. With moist eyes, Maggie hugged them both and said with a catch in her throat, "Never in all my born days have I seen two such beautiful ladies."

Mike cleared his throat and, flushing, said, "Three, I'd say." Then he brusquely waved Maggie away when she thanked him with a kiss.

Mrs. Mole was the first guest to arrive, punctually at four, and she was warmly welcomed by them all. While the three older people decided on a pre-party gossip in the kitchen to catch up on the news at the shelter, Iris and Sarah went outside to report to Grace Ramsey.

The tennis court had been fitted out as a dance floor with four gaily colored canvas tent tops stretched one at each corner to shelter fresh white tables and folding chairs. On the court a small orchestra was setting up and Iris skipped gleefully in a kind of giddy dance to the cheerful, nonsensical sounds of the instruments being tuned up.

"Run on down and watch, if you like, dear," Sarah said, catching sight of Mrs. Ramsey talking to the barmen who'd set up their bar on the north edge of the center lawn. In that position they'd be most easily accessible to the most people, with their backs to the cleanly pruned hedge of oleander that had been such a wretched tangle six months ago.

Grace Ramsey noticed Sarah's approach and she reached out to hug her and kiss her warmly on the cheek before holding Sarah out to be inspected. "How utterly charming you look, Sarah, dear. Those colors are perfect for your hair and your grey eyes. Without a doubt you'll be the most beautiful woman here."

Grace Ramsey wore a pale lavender voile dress of exquisitely simple cut that contrasted dramatically with her thick white hair and brought out the pink glow of her fair skin.

"You look wonderful, too," Sarah said, feeling sad that after today her warm friendship with this lovely woman would be finished. She'd miss the emotional support she'd always received from Mrs. Ramsey and their many quiet talks. During the last six months, Sarah had felt, for the first time in her memory, that she knew what it was to have a mother.

"Sarah, before things get too hectic, I want to thank you for all you've done. In six short months you've turned a wasteland into an Eden. If Ben Yashimoto could see what you've done here, his heart would overflow with pride."

Sarah turned away to conceal the tears in her eyes. Such fulsome praise was almost more than her beleaguered heart could stand right now. In a muffled voice, Sarah replied, "I'm happy that you're satisfied with my work, Mrs. Ramsey, but compared to what you've done for Iris, I've done nothing at all. Without Kight's help, and yours, all of us who love her would've been helpless to save her."

With a little smile, Mrs. Ramsey replied, "Well, Sarah, dear, that's what friends are for." Then, more briskly, she added, "And speaking of the devil, here comes Kight. I'll turn you over to him now, and see to my guests."

Sarah was so moved by his mother's praise that she

noted only in passing how extraordinarily handsome Kight looked today in a cream silk suit and a tan linen shirt open at the neck. He, however, immediately showed a pointed interest in Sarah's attire.

"That's a very pretty dress you're wearing, but I'm surprised that Iris didn't make you wear the dress I brought you." He smiled at her intimately, but as one parent might smile at another.

With a flash of resentment that Kight still thought her in the dark about the true source of his "gift," Sarah said airily, "Actually, the dress doesn't fit me very well. I'm afraid Vivica misjudged my size."

Kight's eyes took on an alert look as he said evenly, "That can't be true, Sarah."

"And why can't it?" she snapped. "I suppose I'm a better judge than you of what fits me and what doesn't."

"Vivica had nothing whatever to do with the selection of either of those dresses." Then with a slanting smile, he added, "And I'll bet my right arm that I know your every measurement down to the inch."

Sarah flushed hotly at this suggestive remark, and to cover her embarrassment, she flared up. "Then you're calling Vivica a liar?"

Kight's eyes narrowed. "No, Sarah, you're calling *me* a liar. What has Vivica to do with this at all?"

Abruptly, Sarah turned her head away, sorry the wretched dress had ever been *made,* let alone purchased by *whichever* of them.

Kight took Sarah's face in his hand and briskly turned it toward him. Frowning, he said, "I think I deserve an answer, Sarah, and I demand to have one."

Refusing to meet his eyes, Sarah mumbled, "She *said* she'd picked it out."

"She actually said that?" he pressed further.

"Well, not in so many words," Sarah admitted, "but there wasn't any doubt—oh, never mind! It doesn't matter, anyway."

"I see," Kight said thoughtfully. Then he added, "As a matter of fact, you're right. It doesn't matter who bought the dress; it wouldn't have been appropriate for today, anyway. After all, it *is* a wedding dress."

Sarah stared at him, her mouth open in surprise. Why hadn't it occurred to her before how unlikely it was that Vivica should pick out for Sarah, of all people, a style of dress once worn for country weddings? But before she could frame a reply, Kight seized her by the elbow and began to walk.

"I'm starving," he said casually. "Come on, let's get something to eat."

Sarah allowed herself to be led, still confused, toward the long buffet table covered with snowy white linen, the focal point in the middle of the lawn. The centerpiece was an enormous cut-glass bowl of succulent pink shrimp mounded over crushed ice. Each end of the table was graced by equally large bowls of jewel-like strawberries from Sarah's own garden. Surrounding each were smaller bowls of sour cream, powdered sugar, and brown sugar, for dipping the strawberries. Sarah smiled proudly to see the perfect shape of the fruit, and judging from the crowds around them, they were as sweet as they looked.

Although she had no appetite, Sarah dutifully put a few pretty canapés and a small crescent of melon wrapped in prosciutto on her plate. Then, as if she had no will of her own, she followed Kight as he sauntered away from the table toward the rose garden.

It was peaceful where they stood, alone for the moment. In the distance Sarah heard the laughter of the swimmers as they dove and frolicked in the pool, and, farther still, the combo's music for the dancers on the court. The crowds of people dressed in whites and creams and summer pastels milled dreamily about on the lawn, laughing, chatting, eating, and drinking, and as Sarah watched them they seemed to her like the large cast of a musical comedy, after everyone's problems have come out right at last, gathered together on the stage for the happy finale. It was a day she'd worked hard for and she'd remember all of it always.

Kight suddenly spoke, breaking into Sarah's reverie. "Too bad Bill isn't here today."

Sarah started, then said uneasily, "Yes, it is. I suppose something important came up to keep him away."

"If you call going to San Diego for the weekend to meet your new lady friend's parents important," Kight said drily.

Sarah flushed a deep red and breathed a gusty sigh of resignation. How horribly embarrassing to have her silly deception exposed like this. What would Kight think of her now? That she was a pathetic fool, of course, a neurotic spinster pretending there was a lover where none existed.

Popping a shrimp into his mouth, Kight said blandly, "All in all, Bill's been a veritable gold mine of information to me. I don't know what I'd have done without him."

A good-looking, well-dressed couple came up to them, and after Kight introduced Sarah to his friends, she made a move to escape. As if it had a life of his own, Kight's hand shot out and gripped Sarah's arm. While the three friends chatted, Sarah stood trapped there, left to stew in her own juices. A busboy wandered by and took their empty plates. Then the couple said good-bye and drifted off.

Kight let go of Sarah's arm and, looking about him with a casual air, remarked, "The garden looks terrific, Sarah. You did a magnificent job. It seems to me—I wonder if you'll agree—wouldn't this be a nice setting for a wedding?"

Sarah's heart plunged. Wasn't it enough that he now knew she had no one? Was it really necessary for her to be subjected to his wedding plans again? She answered in a sharp and brittle voice, "Yes, so Vivica mentioned. She thought the vows should be exchanged by the gazebo there. Has the date been set, then?"

Kight made a great show of peering at his digital wristwatch. "Hmm, I believe at this very moment, the deed is done."

"What?" Sarah said bluntly. "What deed?"

"Vivica is at this very moment a blushing bride in Acapulco."

"No," Sarah flatly contradicted. "She's going to marry you . . . here . . . soon. . . ." But even as she said this, Sarah suddenly realized that she hadn't actually seen Vivica this afternoon, although she'd taken it for granted that she was here and bound to turn up soon to claim her rightful property.

A strange look of what seemed relief passed quickly over Kight's face. "Oh, no, pretty Sarah, *Vivica* isn't the woman who's going to marry me here and soon—and she never has been. Whatever gave you that impression?"

So great was Sarah's indignation at Kight's question that her voice rose as she repeated, "Whatever gave me that impression! Why, *you* did! Vivica said . . . " Sarah blustered, "Your mother did! And it wasn't an impression, it was an *announcement!*"

Kight smiled in an irritating way. "If it was, it must have been similar to the announcement you think Viv made about the dress—that's to say, all in your own head, because nothing could be further from the realm of possibility than marriage between Vivica and me."

Sarah stared at him, her mind turned upside down. Surely she couldn't have misunderstood so completely?

"But you took her along to Mexico with you. . . ."

Kight snorted. "I didn't *take* her, Sarah. Does the ship *take* its barnacles along? I went there on business and Viv just decided to fly down the next day—to use my contacts for a siege of husband-hunting, I think. At least that's how it turned out."

"How do you mean?" Sarah asked, needing to hear such miraculous news spelled out in detail before she dared to believe it.

"She met, and has now married, a very lonely and very grateful silver mine owner. Oh, he's a little older than she, and not in the best of health, but what does that matter to true love?" he said with a lightly sardonic smile. "I think Viv will make him a charming chatelaine in his declining years. At least, she has some experience at it."

Sarah studied his face to detect his true feelings. Was he devastated that Vivica had jilted him twice? Was he pretending unconcern, even amusement, to hide a broken heart? Forgetting herself in her concern for Kight, Sarah hesitantly asked, "But Kight, dear, will you mind . . . I mean, I was so sure you and she—"

Kight put the tip of his index finger on Sarah's lips to stop her words. "I've suspected for some time—I've *hoped*—that you had this wrong side up all along, Sarah, and now I know it's so. Oh, I won't deny that from time to time Viv fools herself into thinking she wants to marry me—mostly when she's bored between love affairs, but she's never taken it seriously. And as for me, I've never felt anything for Viv but what any older brother might feel for a silly pest of a little sister."

"Oh, I see," Sarah breathed, though she still scarcely did see. "Then when you asked me to stay with your mother while you were gone, you didn't know Vivica was coming along? I mean, you *could* have asked *her,* instead?"

Kight made a droll face. "If I'd wanted to risk my mother's disowning me," he joked. "No, Sarah, I asked you because my mother loves you, because I knew I could depend on you, and, most of all—because I love you."

Sarah started, flushing uncomfortably. "Please don't joke about that," she said stiffly.

All amusement slid from Kight's face at her words and he gently drew Sarah into the circle of his arms.

"Why would you think I'd joke about a thing like that, my darling Sarah?" he asked in a husky voice.

Sarah pulled back to look up into his eyes. "You've always teased me, from the first day we met. Why should I believe you now? Bill's out of your way now," she said with a trace of bitterness. "What if this is just one more example of . . . the way of a man with a maid?"

Kight crushed her tighter to him. "Sarah, can you honestly say that's all you thought it was—those times we had together, in the garden, in your house after dinner that evening? Will this convince you it was always more?"

Kight kissed her then and Sarah's lips felt the truth of what her mind found so hard to believe. It was a tender kiss, full of his hungry need, but rich, too, with the commitment of love. When he raised his mouth from hers, Sarah looked deep into his eyes and saw there all she'd ever yearned for in her secret heart.

"Oh, Kight," she cried softly, "why didn't you tell me how you felt?"

Kight held her close and stroked her hair with his hand. "Do you remember that first day we met, you warned me that you'd walk away from any male employer who stepped out of line? Well, I took that warning to heart. As touchy as you were, I was afraid if I made one wrong move I'd lose you forever, especially after I'd made such a damned fool of myself that first day. You were so wary of me, I thought if you worked for my mother instead of me, you'd relax a little and come to trust me."

Sarah began to comprehend how all the painful misunderstandings had begun. "So that's what you meant when you said you'd rather starve than be my employer."

Kight grimaced and laughed shortly. "Yes, and with that feeble attempt at verbal cleverness, I nearly lost you for good. That, and grabbing you the way I did—although I forgive myself for that. I'm not made of iron, after all."

"I think that's where everything went awry," Sarah said thoughtfully. "Because soon after that you saw me with Bill, and I let you think . . ."

"Yes, that's right. Why did you do that, darling?" Kight asked with a sadness that grieved Sarah.

"Oh, I don't even know now!" she cried. "I thought if you thought there was another man, you'd leave me alone."

"So I was right—you *didn't* like me at first," Kight said glumly.

"No, I loved you! I loved you from the first day, but there was Vivica, I thought. And it seemed to me that you were always toying with me. I wouldn't have thought to make that up about Bill, but when you jumped to that conclusion, I just let it go—I thought it would help drive you away." Here Sarah faltered, blushing. "I needed help in resisting you," she said shyly.

With an exultant laugh he pulled her close and kissed her soundly. "Then if it wasn't because of Bill that you so firmly rejected my proposal of marriage that day, and it wasn't because you didn't love me—tell me, Sarah, was it only because of Vivica?"

"Yes, oh, yes, Kight. If you only knew how much I wanted to say yes that day, but I thought you were sacrificing your own happiness for Iris."

Kight caught her close to him and with a catch in his voice, he murmured, "Oh, my red-headed, wrong-headed Sarah. Iris's situation was the reason for choosing that *time*, but *you* were the reason. I wanted you so desperately I was willing to take you any way I could get you—even if I had to steal you from another man."

"Oh, Kight, we've wasted so much time," Sarah said mournfully.

"Never mind. We won't waste any more. Soon we'll be together forever, the three of us at first, then . . ." He bent down and kissed her with sweet, firm lips, and Sarah gave herself up completely in body and spirit.

Kight raised his head and glanced briefly over his shoulder at the garden beyond. In a deep, husky voice, he said, "I think Viv was wrong about the gazebo."

"Mmm?" Sarah murmured, still reveling in the joy of the moment.

"Yes," Kight went on in a comfortable, lazy tone. "We'll take the vows right here, I think, right beside this rose garden, where I first snatched you from the thorns and knew I'd found you at last—my love, who's like a red, red rose, that's newly sprung in June."